Gene Regulation

Gene Regulation

A eukaryotic perspective

Second edition

David Latchman
Department of Molecular Pathology
UCL Medical School
London
UK

CHAPMAN & HALL

London · Glasgow · Weinheim · New York · Tokyo · Melbourne · Madras

Published by Chapman & Hall, 2-6 Boundary Row, London SE1 8HN, UK

Chapman & Hall, 2-6 Boundary Row, London SE1 8HN, UK

Blackie Academic & Professional, Wester Cleddens Road, Bishopbriggs, Glasgow G64 2NZ, UK

Chapman & Hall GmbH, Pappelallee 3, 69469 Weinheim, Germany

Chapman & Hall USA., Fifth Avenue, New York, NY 10003, USA

Chapman & Hall Japan, ITP-Japan, Kyowa Building, 3F, 2-2-1 Hirakawacho, Chiyoda-ku, Tokyo 102, Japan

Chapman & Hall Australia, 102 Dodds Street, South Melbourne, Victoria 3205, Australia

Chapman & Hall India, R. Seshadri, 32 Second Main Road, CIT East, Madras 600 035, India

First edition 1990
Second edition 1995
Reprinted 1995

© 1990, 1995 David Latchman

Typeset in Adobe Garamond by Florencetype Ltd, Stoodleigh, Devon
Printed in Great Britain at the Alden Press, Oxford

ISBN   0 412 60200 8

A Catalogue record for this book is available from the British Library

∞ Printed on permanent acid-free text paper, manufactured in accordance with ANSI/NISO Z39.48-1992 and ANSI/NISO Z39.48-1984(Permanence of Paper).

To my parents

Contents

Preface to the first edition

An understanding of how genes are regulated in humans and higher eukaryotes is essential for the understanding of normal development and disease. For these reasons this process is discussed extensively in many current texts. Unfortunately, however, many of these works devote most of their space to a consideration of gene regulation in bacteria and then discuss the complexities of the eukaryotic situation more briefly. Although such an approach was the only viable one when only the simpler systems were understood, it is clear that sufficient information is now available to discuss eukaryotic gene regulation as a subject in its own right.

This work aims to provide a text which focuses on how cellular genes are regulated in eukaryotes. For this reason bacterial examples are included only to illustrate basic control processes, while examples from the eukaryotic viruses are used only when no comparable cellular gene example of a type of regulation is available.

It is hoped that this approach will appeal to a wide variety of students and others interested in cellular gene regulation. Thus, the first four chapters provide an introduction to the nature of transcriptional regulation as well as a discussion of the other levels at which genes can be regulated, which is suitable for second-year undergraduates in biology or medicine. A more extensive discussion of the details of transcriptional control, at a level suitable for final-year undergraduates, is contained in Chapters 5–7. The entire work should also provide an introduction to the topic for starting researchers or those moving into this area from other fields and wishing to know how the gene which they are studying might be regulated. Clinicians moving into molecular research should find the concentration on the mechanisms of human cellular gene regulation of interest, and for these readers, in particular, Chapter 8 focuses on the malregulation of gene expression which occurs in human diseases and, specifically, in cancer.

I would like to thank Professor Martin Raff and Dr Paul Brickell for their critical reading of the text, and Dr Joan Heaysman who suggested that I should write this book. I am also extremely grateful to Mrs Rose Lang for typing the text and tolerating my continual changes, and to Mrs Jane Templeman who prepared the illustrations.

David S. Latchman

Preface

In the 4 years since the first edition of *Gene Regulation* was published considerable advances have been made in understanding the details of gene regulation. Nonetheless the basic principles outlined in the first edition have remained unchanged and this second edition therefore adopts the same general approach as its predecessor. The opportunity has been taken, however, of updating the discussion of most topics to reflect recent advances. This has required the complete rewriting of some sections where progress has been particularly rapid such as those dealing with alternative splicing factors (Chapter 4) or the mechanism of transcriptional activation (Chapter 7). In addition several new sections have been added reflecting topics which are now known to be of sufficient importance in gene regulation to merit inclusion as well as others suggested by comments on the first edition. These include regulation of transcription by polymerase pausing (Chapter 3), X chromosome inactivation and genomic imprinting (Chapter 5), locus control regions (Chapter 6), negative regulation of transcription (Chapter 7) and anti-oncogenes (Chapter 8). It is hoped that these changes and additions will allow the new edition of *Gene Regulation* to build on the success of the first edition and provide a comprehensive overview of our current understanding in this area.

Finally I would like to thank Mrs Rose Lang who has coped most efficiently with the need to add, delete or amend large sections of the original word processed text of the first edition. I am also most grateful to Rachel Young and the staff at Chapman and Hall for deciding to commission a new edition of this work and their efficiency in producing it.

David S. Latchman

Acknowledgments

I would like to thank all those colleagues who have given permission for material from their papers to be reproduced in this book and have provided prints suitable for reproduction. Other than those listed below, all figures have been drawn especially for this work. Figures 1.4 and 1.5, photographs kindly provided by Dr A. Moore, from Moore *et al., Eur. J. Biochem.* **152**, 729 (1985), by permission of Springer-Verlag; Figure 1.7, redrawn from Bishop *et al., Nature* **250**, 199 (1974), by permission of Dr J. Bishop and Macmillan Magazines Ltd; Figure 1.8, redrawn from Hastie and Bishop, *Cell* **9**, 761 (1976), by permission of Dr J. Bishop and Cell Press; Figure 1.9, from Latchman *et al., Biochim. Biophys. Acta* **783**, 130 (1984), by permission of Elsevier Science Publishers; Figure 1.10, photograph kindly provided by Dr P. Brickell, from Devlin *et al., Development* **103**, 111 (1988), by permission of the Company of Biologists Ltd.

Figure 2.2, photograph kindly provided by Dr A. T. Sumner; Figure 2.4, redrawn from Cove, D. J., *Genetics* (1971), by permission of Professor D. J. Cove and Cambridge University Press; Figure 2.5, redrawn from Steward *et al., Science* **143**, 20 (1964), by permission of Professor F. C. Steward and the American Association for the Advancement of Science (AAAS); Figure 2.6, redrawn from Gurdon, *Gene Expression* (1974), by permission of Professor J. B. Gurdon and Oxford University Press; Figure 2.10, photograph kindly provided by Dr R. Manning, from Manning and Gage, *J. Biol. Chem.* **253**, 2044 (1978); Figure 2.14, redrawn from Spradling and Mahowald, *Proc. Natl. Acad. Sci. USA* **77**, 1096 (1980), by permission of Dr A. Spradling; Figure 2.19, redrawn from Hicks, *Nature* **326**, 445 (1987), by permission of Dr J. Hicks and Macmillan Magazines Ltd.

Figures 3.3 and 3.4, photographs kindly provided by Professor B. W. O'Malley, from Roop *et al., Cell* **15**, 671 (1978), by permission of Cell Press; Figure 3.9, photograph kindly provided by Dr M. Ashburner; Figure 3.11, photograph kindly provided by Dr R. S. Hill, from Hill and Macgregor, *J. Cell Sci.* **44**, 87 (1980), by permission of the Company of Biologists Ltd.

Figure 4.10, redrawn from Breitbart and Nadal-Ginard, *Cell* **49**, 793 (1987), by permission of Professor B. Nadal-Ginard and Cell Press; Figure 4.17, redrawn from Guyette *et al.*, *Cell* **17**, 1013 (1979), by permission of Professor J. Rosen and Cell Press, Figure 4.18, redrawn from Casey *et al.*, *Science* **240**, 924 (1988), by permission of Dr J. B. Harford and AAAS; Figure 4.22, photograph kindly provided by Dr N. Standart and Dr T. Hunt.

Figure 5.3, redrawn from Gurdon, J. B., *Gene Expression* (1974), by permission of Professor J. B. Gurdon and Oxford University Press; Figure 5.4, redrawn from Cove D. J., *Genetics* (1971), by permission of Professor D. J. Cove and Cambridge University Press; Figure 5.6, photographs kindly provided by Dr J. T. Finch, from Finch *et al.*, *Proc. Natl. Acad. Sci. USA,* **72**, 3320 (1975); Figure 5.7, photograph kindly provided by Dr F. Thoma, from Thoma *et al.*, *J. Cell. Biol.* **83**, 403 (1979), by permission of Rockefeller Press; Figure 5.8, photograph kindly provided by Dr J. McGhee, from McGhee *et al.*, *Cell* **33**, 831 (1983), by permission of Cell Press; Figure 5.10, photograph kindly provided by Professor O. L. Miller, from McKnight *et al.*, *Cold Spring Harbor Symp.* **42**, 741 (1978), by permission of Cold Spring Harbor Laboratory; Figure 5.15, redrawn from Weintraub *et al.*, *Cell* **24**, 333 (1981), by permission of Professor H. Weintraub and Cell Press; Figure 5.19, redrawn from Keshet *et al.*, *Cell* **44**, 535 (1986), by permission of Professor H. Cedar and Cell Press; Figure 5.23, photograph kindly provided by Professor P. Chambon, from Kaye *et al.*, *EMBO J.* **3**, 1137 (1984), by permission of Oxford University Press; Figure 5.25, redrawn from Noreheim and Rich, *Nature* **303**, 674 (1983), by permission of Professor A. Rich and Macmillan Magazines Ltd; Figure 5.26, photograph kindly provided by Professor M. Yaniv, from Saragosti *et al.*, *Cell* **20**, 65 (1980), by permission of Cell Press.

Figures 6.12 and 6.13, photographs kindly provided by Professor D. Hanahan, from Hanahan, *Nature* **315**, 115 (1985), by permission of Macmillan Magazines Ltd; Figures 6.22 and 6.24, photographs kindly provided by Professor D. D. Brown, 6.22 from Sakonju *et al.*, *Cell* **19**, 13 (1980), 6.24 from Sakonju and Brown, *Cell* **31**, 395 (1982), by permission of Cell Press.

Figure 7.4, photograph kindly provided by Professor W. J. Gehring, from Gehring, *Science* **236**, 1245 (1987), by permission of AAAS; Figure 7.10, redrawn from Schleif, *Science* **241**, 1182 (1988), by permission of Dr R. Schleif and AAAS; Figure 7.12, redrawn from Evans and Hollenberg, *Cell* **52**, 1 (1988), by permission of Professor R. M. Evans and Cell Press; Figure 7.13, redrawn from Klug and Rhodes, *Trends in Biochemical Science* **12**, 464 (1987), by permission of Professor Sir Aaron Klug and

Elsevier Publications; Figure 7.21, redrawn from Turner and Tjian, *Science* **243**, 1189 (1989), by permission of Professor R. Tjian and AAAS; Figure 8.3, redrawn from Takeya and Hanafusa, *Cell* **32**, 881 (1985), by permission of Professor H. Hanafusa and Cell Press.

Tissue-specific expression of proteins and messenger RNAs

1

1.1 Introduction

The evidence that eukaryotic gene expression must be a highly regulated process is available to anyone visiting a butcher's shop. The various parts of the mammalian body on display differ dramatically in appearance, ranging from the muscular legs and hind quarters to the soft tissues of the kidneys and liver. However, all these diverse types of tissues arose from a single cell, the fertilized egg or zygote, raising the question of how this diversity is achieved. It is the aim of this book to consider the processes regulating tissue-specific gene expression in mammals and the manner in which they produce these differences in the nature and function of different tissues.

1.2 Tissue-specific expression of proteins

The fundamental dogma of molecular biology is that DNA produces RNA which in turn produces proteins. Thus the genetic information in the DNA specifying particular functions is converted into an RNA copy, which is then translated into protein. The action of the protein then produces the phenotype, be it the presence of a functional globin protein transporting oxygen in the blood or the activity of a proteinaceous enzyme capable of producing the pigment causing the appearance of brown rather than blue eyes. Hence, if the differences in the appearance of mammalian tissues described above are indeed caused by differences in gene expression in different tissues, they should be produced by differences in the proteins present in these tissues. Such differences can be detected both by general methods aimed at studying the expression of all proteins in a given tissue

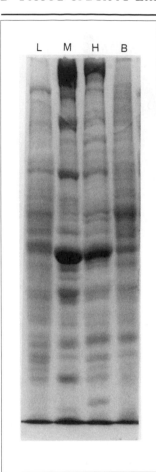

Figure 1.1 One-dimensional SDS–polyacrylamide gel electrophoresis of cellular proteins isolated from mouse brain (track B), mouse heart (track H), mouse liver (track L) and mouse skeletal muscle (track M).

and by specific methods which study the expression of one particular protein.

1.2.1 General methods for studying the protein composition of tissues

If the proteins in a cell are denatured by treatment with the detergent sodium dodecyl sulfate (SDS) and then subjected to electrophoresis in a polyacrylamide gel (Laemmli, 1970), they can be visualized by staining the gel. This method allows the demonstration of some differences between different cells and tissues (Figure 1.1). Because of the very large number of proteins in the cell and the limited resolution of the technique, one-dimensional gel electrophoresis cannot be used to extensively investigate the variation of proteins between tissues. Thus, for example, two entirely different proteins in two tissues may be scored as being the same protein simply on the basis of a similarity in size. A more detailed investigation of the protein composition of different tissues can be achieved by two-dimensional gel electrophoresis (O'Farrell, 1975). In this procedure (Figure 1.2), proteins are first separated on the basis of differences in their charge, in a technique known as isoelectric focusing, and the separated proteins, still in the first gel, are layered on top of an SDS–polyacrylamide gel. In the subsequent electrophoresis the proteins are separated by their size. Hence a protein moves to a position determined both by its size and its charge. The much greater resolution of this method allows a number of differences in the protein composition of particular tissues to be identified. Thus some spots or proteins are found in only one or a few tissues and not in many others, while others are found at dramatically different abundance in different tissues (Figure 1.3). Hence, the different appearances of different tissues are indeed paralleled by both qualitative and quantitative differences in the proteins present in each tissue. It should be noted, however, that some proteins can be shown by two-dimensional gel electrophoresis to be present at similar levels in virtually all tissues. Presumably such so-called housekeeping proteins are involved in basic metabolic processes common to all cell types.

1.2.2 Specific methods for studying the protein composition of tissues

The general conclusion that qualitative and quantitative differences exist in the protein composition of different tissues can also be reached by investigating the expression of particular individual proteins in such tissues. These methods might involve the isolation of widely differing amounts of an individual protein (or none of the protein at all) from

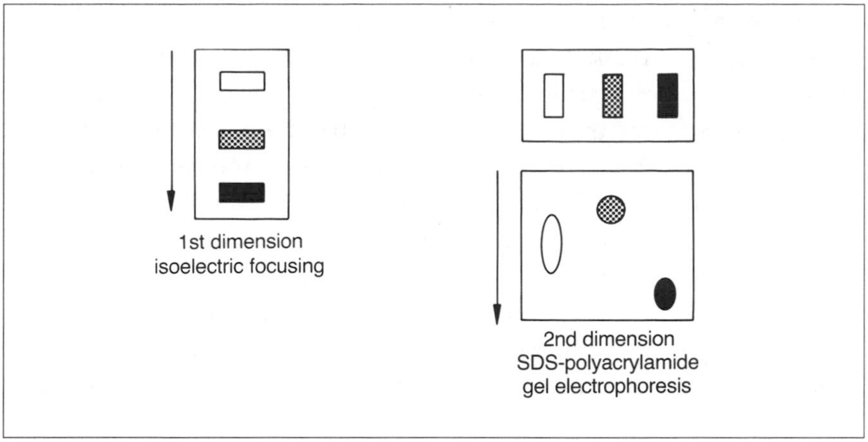

Figure 1.2 Two-dimensional gel electrophoresis.

1st dimension
isoelectric focusing

2nd dimension
SDS-polyacrylamide
gel electrophoresis

different tissues, using an established purification procedure, or the detection of an enzymatic activity associated with the protein in extracts of only one particular tissue. We shall consider in detail, however, only methods where the expression of a specific protein is monitored by the use of a specific antibody to it. Normally, such an antibody is produced by injecting the protein into an animal such as a rabbit or mouse. The resulting immune response results in the presence in the animal's blood of antibodies which specifically recognize the protein and can be used to monitor its expression in particular tissues.

Such antibodies can be used in conjunction with the one-dimensional polyacrylamide gel electrophoresis technique already described (Section 1.2.1) to investigate the expression of a particular protein in different tissues. In this technique, known as Western blotting (Gershoni and Palade, 1983), the gel-separated proteins are transferred to a nitrocellulose filter which is incubated with the antibody. The antibody reacts specifically with the protein against which it is directed and which will be present

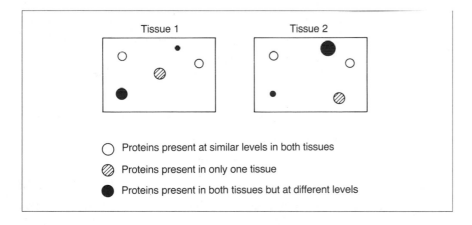

Tissue 1   Tissue 2

○ Proteins present at similar levels in both tissues

⊘ Proteins present in only one tissue

● Proteins present in both tissues but at different levels

Figure 1.3 Schematic results of two-dimensional gel electrophoresis allowing the detection of proteins specific to one tissue or expressed at different levels in different tissues.

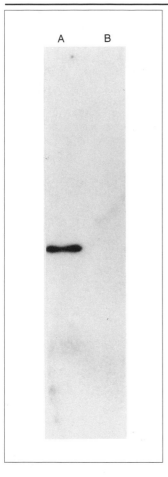

Figure 1.4 Western blot with an antibody to guinea-pig casein kinase showing the presence of the protein in lactating mammary gland (A) but not in liver (B).

at a particular position on the filter, dependent on how far it moved in the electrophoresis step and hence on its size. The binding of the antibody is then visualized by a radioactive or enzymatic detection procedure. If a tissue contains the protein of interest, a band will be observed in the track containing total protein from that tissue and the intensity of the band observed will provide a measure of the amount of protein present in the tissue. If none of the particular protein is present in a given tissue, no band will form (Figure 1.4). Hence this method allows the presence or absence of a specific protein in a particular tissue to be assessed using one-dimensional gel electrophoresis without the complicating effect of other unrelated proteins of similar size, since these will fail to bind the antibody. Similarly, quantitative differences in the expression of a particular protein in different tissues can be detected on the basis of differences in the intensity of the band obtained when extracts prepared from the different tissues are used in this procedure.

The specific reaction of a protein with an antibody can also be used directly to investigate its expression within a particular tissue. In this method, known as immunofluorescence, thin sections of the tissue of interest are reacted with the antibody, which binds to those cell types expressing the protein. As before, the position of the antibody is visualized by an enzymatic detection procedure, or more usually, by the use of a fluorescent dye which can be seen in a microscope when appropriate filters are used (Figure 1.5). Therefore not only can this method be used to provide information about the tissues expressing a particular protein but, since individual cells can be examined, it also allows detection of the specific cell types within the tissue which are expressing the protein.

As with the general methods for studying all proteins, specific methods aimed at studying the expression of particular proteins indicate that while some proteins are present at similar abundance in all tissues, others are present at widely different abundances in different tissues and a large number are specific to one or a few tissues or cell types. Hence the differences between different tissues in appearance and function are correlated with qualitative and quantitative differences in protein composition, and it is necessary to understand how such differences are produced.

1.3 Tissue-specific expression of messenger RNAs

Proteins are produced by the translation of specific mRNA molecules on the ribosome. Hence, having established that quantitative and qualitative

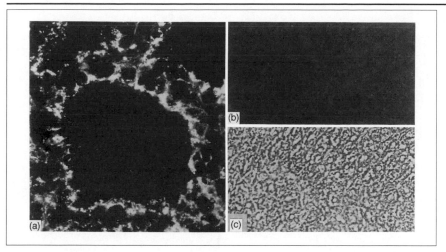

Figure 1.5 Use of the antibody to casein kinase to show the presence of the protein (bright areas) in frozen sections of lactating mammary gland (a) but not in liver (b). (c) A phase-contrast photomicrograph of the liver section, confirming that the lack of staining with the antibody in (b) is not due to the absence of liver cells in the sample.

differences exist in the protein composition of different tissues, it is necessary to ask whether such differences are paralleled by tissue-specific differences in the abundance of their corresponding mRNAs. Thus, although it seems likely that differences in the mRNA populations of different tissues do indeed underlie the observed differences in proteins, it is possible that all tissues have the same mRNA species and that production of different proteins is controlled by regulating which of these are selected by the ribosome to be translated into protein.

As with the study of proteins, both general and specific techniques exist for studying the mRNAs expressed in a given tissue.

1.3.1 General methods for studying the mRNAs expressed in different tissues

The most commonly used method for studying RNA populations is the $R_0 t$ curve (Bishop *et al.*, 1974). In this method a complementary DNA (cDNA copy) of the RNA is first made, using radioactive precursors so that the cDNA is made radioactively labeled. After dissociating the cDNA and the RNA, the rate at which the radioactive cDNA reanneals or hybridizes to the RNA is followed. In this process, more molecules of an abundant RNA will be present than of an RNA which is less abundant. The cDNA from an abundant RNA will find a partner more rapidly than one derived from a rare one, simply because more potential partners are available. Hence, when used with the RNA of one tissue, the rate of hybridization or reannealing of any cDNA molecule provides a measure of the abundance of its corresponding RNA in that tissue. The higher the abundance of a particular RNA species, the more rapidly it anneals to its cDNA (Figure 1.6). Typical $R_0 t$ curves, showing three

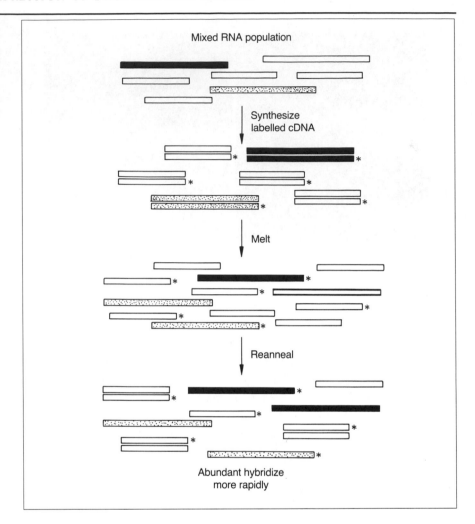

Figure 1.6 Hybridization of mRNA and its radioactive cDNA copy (indicated by the stars) in a $R_o t$ curve experiment. The abundant RNA finds its partner and hybridizes more rapidly.

predominant abundance classes of RNA in any given tissue, can be constructed readily (Figure 1.7).

This method can be extended to provide a comparison of the RNA populations of two different tissues (Hastie and Bishop, 1976). In this method the RNA of one tissue is mixed with labeled cDNA prepared from the RNA of another tissue and the mixture annealed. Clearly, the rate of hybridization of the radiolabeled cDNA will be determined by the abundance of its corresponding RNA in the tissue from which the RNA is derived. If the RNA is absent altogether in this tissue (being specific to that from which the cDNA is derived), no reannealing will occur. If the corresponding RNA is present, the rate at which the cDNA hybridizes will be determined by its abundance in the tissue from which the RNA is derived. Hence, by carrying out an inter-tissue $R_o t$ curve, both qualitative

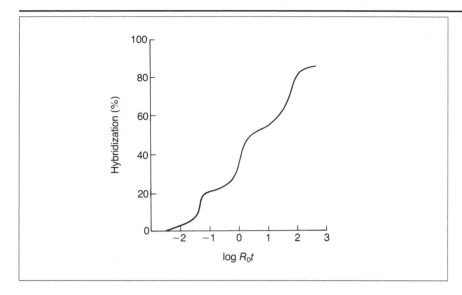

Figure 1.7 Typical $R_o t$ curve, produced using the mRNA of a single tissue. Note the three phases of the curve, produced by hybridization of the highly abundant, moderately abundant and rare RNA species.

and quantitative differences between the RNA populations of different tissues can be observed. Such experiments (Figure 1.8) lead to the conclusion that, although sharing of rare RNAs (presumably encoding housekeeping proteins) does occur, both qualitative and quantitative differences between the RNA populations of different tissues are observed, with RNA species that are abundant in one cell type being rare or absent in other tissues. Thus, in a comparison of kidney and liver mRNAs, Hastie and Bishop (1976) found that while between 9500 and 10 500 of the 11 000 rare mRNA species found in kidney were also found in liver, the six highly

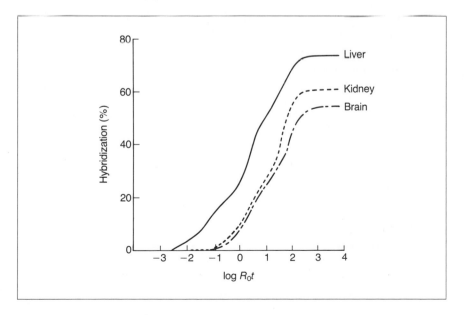

Figure 1.8 Inter-tissue $R_o t$ curve, in which mRNA from the liver has been hybridized to cDNA prepared from liver, kidney or brain mRNA. Note the lower extent of hybridization, with the kidney and brain samples indicating the absence or lower abundance of some liver RNAs in these tissues.

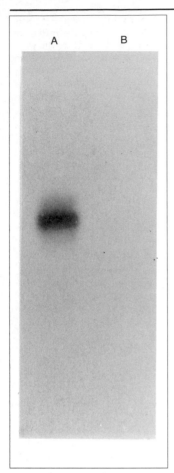

Figure 1.9 Northern blot hybridization using a probe specific for the α-fetoprotein mRNA. The RNA is detectable in the embryonic yolk sac sample (track A) but not in the adult liver sample (track B).

abundant RNA species found in the kidney were either absent or present at very low levels in the liver.

1.3.2 Specific methods for studying the mRNAs expressed in different tissues

The conclusion from $R_o t$ curve analysis, that both qualitative and quantitative differences exist between RNA populations, can also be reached by the use of specific methods that detect one specific mRNA using a cloned DNA probe derived from its corresponding gene. In the most commonly used of such methods, Northern blotting (Thomas, 1980), the RNA extracted from a particular tissue is electrophoresed on an agarose gel, transferred to a nitrocellulose filter and hybridized to a radioactive probe derived from the gene encoding the mRNA of interest. The presence of the RNA in a particular tissue will result in binding of the radioactive probe and the visualization of a band on autoradiography, the intensity of the band being dependent on the amount of RNA present (Figure 1.9). Although such experiments do detect the RNA encoding some proteins, such as actin or tubulin, in all tissues, very many others are found only in one particular tissue. Thus the RNA for the globin protein is found only in reticulocytes, that for myosin only in muscle, while (in the example shown in Figure 1.9) the mRNA encoding the fetal protein, α-fetoprotein, is shown to be present only in the embryonic yolk sac and not in the adult liver.

As with protein studies, methods studying expression in RNA isolated from particular tissues can be supplemented by methods allowing direct visualization of the RNA in particular cell types. In such a method, known as *in situ* hybridization, a radioactive probe specific for the RNA to be detected is hybridized to a section of the tissue of interest in which cellular morphology has been maintained. Visualization of the position at which the radioactive probe has bound (Figure 1.10) allows an assessment not only of whether the particular tissue is expressing the RNA of interest but also of which individual cell types within the tissue are responsible for such expression. This technique has been used to show that the expression of some mRNAs is confined to only one cell type, paralleling the expression of the corresponding proteins by that cell type.

1.4 Conclusions

It is clear, therefore, that the qualitative and quantitative variation in protein composition between different tissues is paralleled by a similar

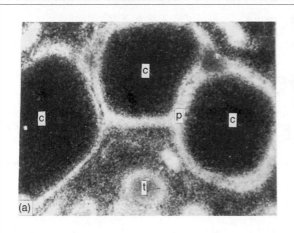
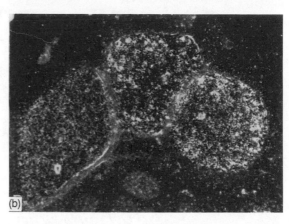

variation in the nature of the mRNA species present in different tissues. Hence, the production of different proteins by different tissues is not regulated by selecting which of a common pool of RNA molecules is selected for translation by the ribosome, although some isolated examples of such regulated translation may exist (Chapter 4). In order to understand the basis for such tissue-specific variation in protein and mRNA composition, it is therefore necessary to move one stage further back and examine the nature of the DNA encoding these mRNAs in different tissues.

Figure 1.10 Localization of the RNA for type 1 collagen (A) and type II collagen (B) in the 10 day chick embryo leg by *in situ* hybridization. Note the different distributions of the bright areas produced by binding of each probe to its specific mRNA. c, cartilage; t, tendons; p, perichondrium.

References

Bishop, J. O., J. G. Morton, M. Rosbash and M. Richardson, 1974. Three abundance classes in HeLa cell messenger RNA. *Nature* **250**, 199–204.

Gershoni, J. M. and G. E. Palade, 1983. Protein blotting: principles and applications. *Analytical Biochemistry* **131**, 1–15.

Hastie, N. B. and J. O. Bishop, 1976. The expression of three abundance classes of messenger RNA in mouse tissues. *Cell* **9**, 761–74.

Laemmli, U. K. 1970. Cleavage of structural proteins during the assembly of the head of the bacteriophage T4. *Nature* **227**, 680–5.

O'Farrell, P. H. 1975. High resolution two-dimensional electrophoresis of proteins. *Journal of Biological Chemistry* **250**, 4007–21.

Thomas, P. S. 1980. Hybridization of denatured RNA and small DNA fragments transferred to nitrocellulose. *Proceedings of the National Academy of Sciences of the USA* **77**, 5201–5.

2 The DNA of different cell types is similar in both amount and type

2.1 Introduction

Having established that the RNA and protein content of different cell types shows considerable variation, the fundamental dogma of molecular biology, in which DNA makes RNA which in turn makes protein, directs us to consider whether such variation is caused by differences in the DNA present in each tissue. Thus, in theory, DNA which corresponded to an RNA required in one particular tissue only might be discarded in all other tissues.

Alternatively, it might be activated in the tissue where RNA was required, by a selective increase in its copy number in the genome via an amplification event or by some rearrangement of the DNA necessary for its activation. The possible use of each of these mechanisms, DNA loss, amplification or rearrangement, will be discussed in turn (reviewed by Brown, 1981).

2.2 DNA loss

2.2.1 DNA loss as a mechanism of gene regulation

In theory, a possible method of gene control would involve the genes for all proteins being present in the fertilized egg, followed by a selective loss of genes which were not required in particular tissues (Figure 2.1). Thus only the genes for housekeeping proteins and the proteins characteristic of a particular cell type would be retained in that cell type. Hence the presence of the RNA for immunoglobulin only in the antibody-producing B cells would be paralleled by the presence of the corresponding gene only in such cells, whereas the gene for a muscle-specific protein (such as

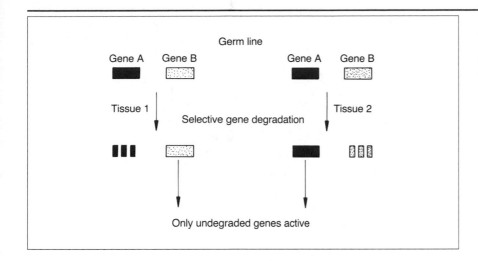

Germ line

Gene A Gene B   Gene A Gene B

Tissue 1 | Selective gene degradation | Tissue 2

Only undegraded genes active

Figure 2.1 Model for gene regulation by the specific degradation of genes whose products are not required in a particular tissue.

myosin) would be absent in B cell DNA but present in that of a muscle cell, where the immunoglobulin gene had been deleted. Such a mechanism would obviate the need for any other means of regulating gene activity, the tissue-specific RNA and protein populations described in Chapter 1 being produced by the constitutive activity of the genes retained in any particular cell type.

A variety of evidence suggests, however, that selective DNA loss is not a general mechanism of gene control. Such evidence is available from chromosomal, functional and molecular studies, which will be considered in turn.

2.2.2 Chromosomal studies

The DNA containing chromosomes, which can be visualized shortly before cell division, differ from one another in both size and the location of the centromere, by which they attach to the mitotic spindle prior to cell division. They can be further distinguished by a variety of staining techniques, which produce a specific banded appearance. The most widely used of these, known as G banding, involves brief proteolytic or chemical treatment of the chromosome, followed by staining with Giemsa stain, and produces a characteristic pattern of bands (Figure 2.2) allowing the cytogeneticist to identify individual chromosomes in humans and other species and to distinguish cells from different species.

In general, when such techniques are applied to chromosomes prepared from different differentiated cell types of the same species, however, the numbers and sizes of the chromosomes observed and their banding patterns are indistinguishable. Thus the patterns obtained from a liver cell or a brain cell are very similar and cannot be distinguished even by

Figure 2.2 Human chromosomes stained with Giemsa stain. The upper panel shows a chromosome spread; in the lower panel the chromosomes have been arrayed in order of their chromosome number.

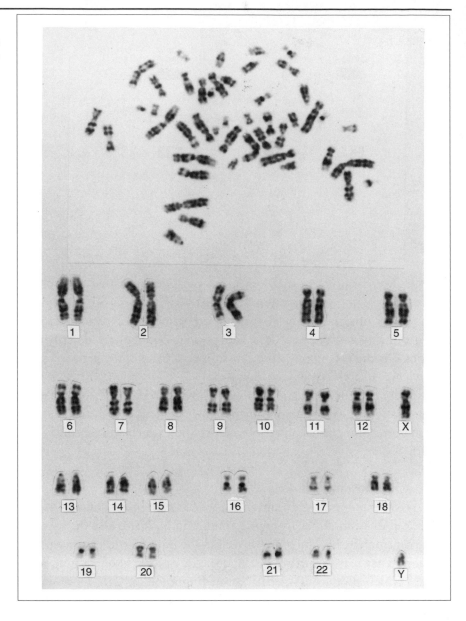

experienced cytogeneticists. Hence such karyotypic studies provide strong evidence against the loss of whole chromosomes or specific parts of chromosomes (visualized as G bands) in particular differentiated cell types and argue against selective DNA loss as a general mechanism for gene regulation.

None the less, a small number of cases of such a loss of DNA have been observed in particular cell types and these will be considered in turn.

Loss of the nucleus in mammalian red blood cells

The best-known case occurs in the development of the mammalian red blood cell, or erythrocyte (for review, see Bessis and Brickal, 1952), which is a highly specialized cell containing large amounts of the blood pigment hemoglobin and functions in the transport of oxygen in the blood. Such cells are produced from progenitors, known as erythroblasts, which divide to produce two daughter cells, one of which maintains the stem cell population and later divides again, while the other undergoes differentiation. In this differentiation, the region of the cell containing the nucleus is pinched off, surrounded by a region of the cell membrane and eventually destroyed (Figure 2.3). The resulting cell, known as the reticulocyte, is thus entirely anucleate and completely lacks DNA. However, it continues synthesis of large amounts of globin (the protein part of the hemoglobin molecule) as well as small amounts of other reticulocyte proteins, by repeated translation of mRNA molecules produced prior to the loss of the nucleus. Such RNA molecules must therefore be highly stable and resistant to degradation (Chapter 4). Eventually, however, other cytoplasmic components (including ribosomes) are lost, protein synthesis ceases and the cell assumes the characteristic structure of the erythrocyte, which is essentially a bag full of oxygen-transporting hemoglobin molecules.

Although this process offers us an example of DNA loss during differentiation, it is clearly a highly specialized case in which the loss of the nucleus is primarily intended to allow the cell to fill up with hemoglobin and to assume a shape facilitating oxygen uptake, rather than as a means of gene control. Thus the genes for globin and other reticulocyte proteins are not selectively retained but are lost with the rest of the nuclear DNA. Hence the tissue-specific pattern of protein synthesis observed in the reticulocyte is achieved not by selective gene loss but by the processes which, earlier in development, resulted in high-level transcription of the globin genes and the production of long-lived, stable mRNA molecules.

Loss of specific chromosomes in somatic cells

The reticulocyte case does not, therefore, provide us with an example of selective DNA loss and, indeed, no such case has as yet been reported in mammalian cells. However, a number of such cases in which specific chromosomes are lost during development have been described in other eukaryotes, including some nematodes, crustaceans and insects (Wilson, 1928). In these cases a clear distinction is made early in embryonic development between the cells which will produce the body of the organism (the somatic cells) and those which will eventually form its gonads and allow the organism to reproduce (the germ cells). Early on in the

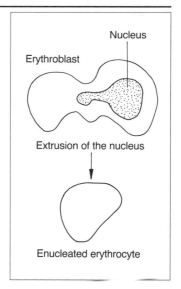

Figure 2.3 Extrusion of the nucleus from an erythroblast, resulting in an anucleate erythrocyte.

development of the embryo, during the cleavage stage, the somatic cells lose certain chromosomes, with only the germ cells, which will give rise to subsequent generations, retaining the entire genome. Such losses can be quite extensive; for example, in the gall midge *Miastor* only 12 of the total 48 chromosomes are retained in the somatic cells. It is assumed that the lost DNA is only required for some aspects of germ cell development. It is known to be rich in the highly repeated (or satellite) type of DNA, which is thought to be involved in homologous chromosome pairing during the meiotic division that occurs in germ cell development (John and Miklos, 1979). Hence it is more economical for the organism to dispense with such DNA in the development of somatic tissue, rather than to waste energy replicating it and passing it on to all somatic cells where it has no function.

Although the mechanism by which this selective loss of DNA occurs is not yet fully understood, there is evidence that it is mediated by cytoplasmic differences between regions of the fertilized egg or zygote. Thus, in the nematode worm *Parascaris* the distinction between somatic and germ cells is established by the first division of the zygote, the smaller of the two cells produced losing chromosomes and giving rise to the soma while the larger retains a full chromosome complement and gives rise to the germ line (Figure 2.4). If, however, the zygote is centrifuged, the plane of the first division can be altered, resulting in the production of two similar-sized cells neither of which loses chromosomes. Hence the centrifugation

Figure 2.4 Early development in the nematode and the consequences of changing the plane of the first cell division by centrifugation of the zygote.

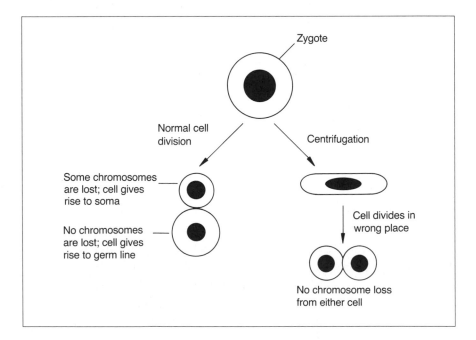

process has abolished some heterogeneity in the cytoplasm responsible for this effect. Interestingly, although both daughter cells retain a full complement of chromosomes, their inability to lose chromosomes prevents normal embryonic development from occurring.

Although cases of DNA loss observable by examination of the chromosomes have been observed in some organisms, it must be emphasized that in the vast majority of cases no such loss of DNA is observed by this means. However, because the amount of DNA in a Giemsa-stained band is very large (about 10 million base pairs), it is possible that losses of short regulatory regions necessary for gene expression, or losses of individual genes, could occur and not be observed by this technique. The large-scale losses seen in chromosome studies would then represent the tip of the iceberg in which varying degrees of gene loss occur in different situations. A consideration of this possibility requires the use of more sensitive methods to search for gene deletions. These involve both functional and molecular studies.

2.2.3 Functional studies

A model in which selective gene loss controls differentiation must view differentiation as essentially an irreversible process. Thus, for example, once the gene for myosin has been eliminated from an antibody-producing B lymphocyte it will not be possible for such a cell, whatever the circumstances, to give rise to a myosin-containing muscle cell. Although the requirement for such a change from one differentiated cell type to another is not likely to occur in normal development, such a phenomenon, known as transdifferentiation, has been achieved experimentally in Amphibia (Yamada, 1967). Thus, if the lens is surgically removed from an eye of one of these organisms, some of the neighboring cells in the iris epithelium lose their differentiated phenotype and begin to proliferate. Eventually these cells differentiate into typical lens cells and produce large quantities of the lens-specific proteins, such as the crystallins, which are readily detectable with appropriate antibodies. The genes for the lens proteins must therefore be intact within the iris cells even though in normal development such genes would never be required in these cells.

The case against selective loss of DNA is strengthened still further by the very dramatic experiments in which a whole new organism has been produced from a single differentiated cell. Although not yet achieved in mammalian cells, this has been achieved in both plants (Steward, 1970) and Amphibia (Gurdon, 1968).

In plants such regeneration has been achieved in several species, in-

Figure 2.5 Scheme for the production of a fertile carrot plant from a single differentiated cell of the adult plant phloem, by growth in culture as a free cell suspension which develops into an embryo and then into an adult plant.

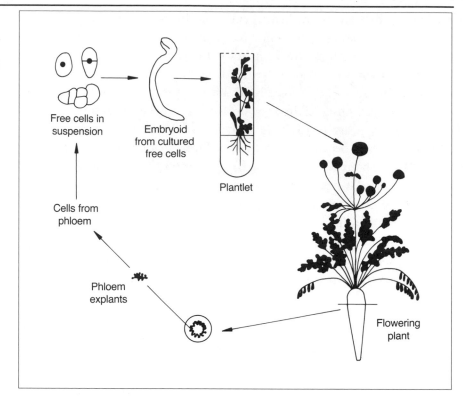

cluding both the carrot and tobacco plants. In the carrot, for example, regeneration can occur from a single differentiated phloem cell, which forms part of the tubing system by which nutrients are transported in the plant. Thus if a piece of root tissue is placed in culture (Figure 2.5), single quiescent cells of the phloem can be stimulated to grow and divide, and an undifferentiated callus-type tissue forms. The disorganized cell mass can be maintained in culture indefinitely but if the medium is suitably supplemented at various stages, embryonic development will occur and eventually result in a fully functional flowering plant, containing all the types of differentiated cells and tissues normally found. The plant that forms is fertile and cannot be distinguished from a plant produced by normal biological processes.

The ability to regenerate a functional plant from fully differentiated cells eliminates the possibility that genes required in other cell types are eliminated in the course of plant development. It is noteworthy, however, that in this case, as with lens regeneration, one differentiated cell type does not transmute directly into another; rather a transitional state of undifferentiated proliferating cells serves as an intermediate. Hence although differentiation does not apparently involve permanent irreversible changes in the DNA, it appears to be relatively stable and, although reversible,

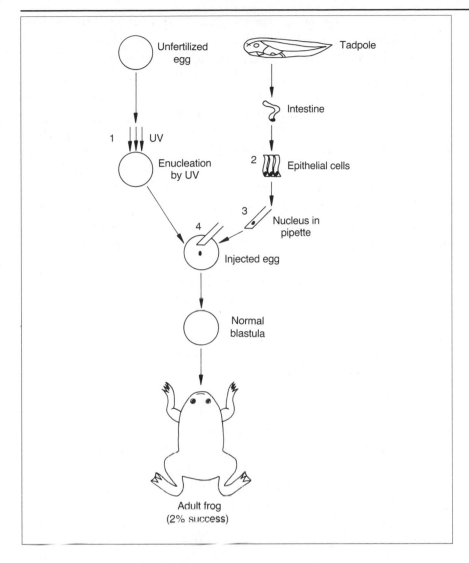

Unfertilized egg

Tadpole

Intestine

1 UV

Enucleation by UV

2 Epithelial cells

3 Nucleus in pipette

4 Injected egg

Normal blastula

Adult frog (2% success)

Figure 2.6 Nuclear transplantation in Amphibia. The introduction of a donor nucleus from a differentiated cell into a recipient egg whose nucleus has been destroyed by irradiation with ultraviolet light can result in an adult frog with the genetic characteristics of the donor nucleus.

requires an intermediate stage for changes to occur. This semi-stability of cellular differentiation will be discussed further in Chapter 5.

Although no complex animal has been regenerated by culturing a single differentiated cell in the manner used for plants, other techniques have been used to show that differentiated cell nuclei are capable of giving rise to very many different cell types (Gurdon, 1968). These experiments involve the use of nuclear transplantation (Figure 2.6). In this technique the nucleus of an unfertilized frog egg is destroyed, either surgically or by irradiation with ultraviolet light, and a donor nucleus from a differentiated cell of a genetically distinguishable strain of frog is implanted. Development is then allowed to proceed in order to test whether, in the environment of the egg cytoplasm, the genetic information in the nucleus of the differentiated cell can produce an adult frog. When the donor nucleus is derived, e.g. from differentiated intestinal epithelial cells of a tadpole, an adult frog is indeed produced in a small proportion of cases (about 2%) and development to at least a normal swimming tadpole occurs in about 20% of cases, with the organisms having the genetic characteristics of the donor nucleus. Such successful development supported by the nucleus of a differentiated frog cell is not unique to intestinal cells and has been achieved with the nuclei of other cell types, such as the skin cells of an adult frog. Hence although these experiments are technically difficult and have a high failure rate, it is possible for the nuclei of differentiated cells specialized for the absorption of food to produce organisms containing a range of different cell types and which are perfectly normal and fertile.

As with plants, it is clear that in animals selective loss of DNA is not a general mechanism by which gene control is achieved.

2.2.4 Molecular studies

The conclusions of functional studies carried out in the 1960s and early 1970s were confirmed abundantly in the late 1970s and 1980s by molecular investigations involving the use of specific DNA probes derived from individual genes. Such probes can be used in Southern blotting experiments (Southern, 1975) to investigate the structure of a particular gene in different tissues. In this technique (Figure 2.7) DNA from a particular tissue is digested into specific fragments with restriction endonucleases which cut the DNA at particular points. Following electrophoresis on an agarose gel, the cut DNA is transferred to nitrocellulose and hybridized with a radioactive probe for the gene being studied. The structure of the gene can then be examined by determining the sizes of fragments observed when the DNA is cut with particular restriction enzymes.

As with the use of Northern blotting and Western blotting (Chapter 1)

Figure 2.7 Procedure for Southern blot analysis, involving electrophoresis of DNA which has been cut with a restriction enzyme, its transfer to a nitrocellulose filter and hybridization of the filter with a specific DNA probe for the gene of interest.

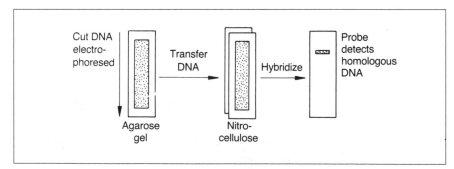

to study, respectively, the RNA and protein content of different tissues, Southern blotting can be used to investigate the structure of the DNA in individual tissues. When this is done, no losses of particular genes are observed in specific tissues. Thus, the gene for β-globin is clearly present in the DNA of brain or spleen where it is never expressed, as well as in the DNA of erythroid tissues where expression occurs (Jeffreys and Flavell, 1977) and numerous other examples have been reported (Figure 2.8).

These techniques, which could detect losses of less than 100 bases (as opposed to the millions of bases resolvable by chromosomal techniques), have now been supplemented by cases where different laboratories have isolated particular genes from the DNA of different tissues which either do or do not express the gene. When the DNA sequence of the gene was determined in each case, no differences were detected in the expressing or non-expressing tissues. Such techniques, which could detect the loss of even a single base of DNA, have therefore confirmed the conclusions of all other studies, indicating that selective DNA loss is not a general mechanism of gene regulation.

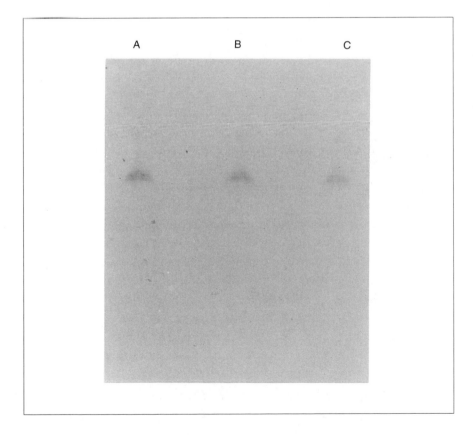

Figure 2.8 Southern blot hybridization of a radioactive probe specific for the α-fetoprotein gene to *Eco*RI-cut DNA prepared from fetal liver (track A), adult liver (track B) and adult brain (track C). Note the presence of an identically sized gene fragment in each tissue, although the gene is only expressed in the fetal liver.

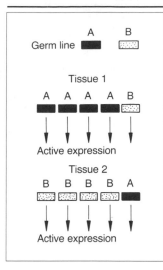

Figure 2.9 Model for gene regulation in which a gene that is expressed at high levels in a particular tissue is selectively amplified in that tissue. Gene A is amplified in tissue 1 and gene B in tissue 2.

2.3 *DNA amplification*

2.3.1 DNA amplification as a mechanism of gene regulation

Having eliminated selective DNA loss as a general means of gene control, it is necessary to consider the possible role of gene amplification in the regulation of gene expression (for reviews, see Stark and Wahl, 1984; Kafatos *et al.*, 1985). A possible mechanism of gene control (Figure 2.9) would involve the selective amplification of genes that were expressed at high levels in a particular tissue, the high expression of such genes simply resulting from normal rates of transcription of the multiple copies of the gene. Unlike gene-deletion models, this possibility is not incompatible with the ability of nuclei from differentiated cells, such as frog intestinal cells, to regenerate a new organism (Section 2.2.3). Thus, although amplification of intestine-specific genes would have occurred in such cells, this would presumably not prevent the amplification of other tissue-specific genes in the individual cell types of the new organism. The evidence against this idea comes therefore from chromosomal and molecular rather than functional investigations.

2.3.2 Chromosomal studies

As with DNA loss, the identity of the chromosomal complement in individual cell types provides strong evidence against the occurrence of large-scale DNA amplification. None the less, some cases of such amplification have been observed by this means and these will be discussed.

Chromosome polytenization in Drosophila

One such case, which has been extensively used in cytogenetic studies, occurs in the salivary gland cells of the fruit fly *Drosophila melanogaster*. In these cells the DNA replicates repeatedly and the daughter molecules do not separate but remain close together, forming a single giant, or polytene, chromosome (Figure 2.10) which contains approximately 1000 DNA molecules. Although such giant chromosomes have proved very useful in experiments involving visualization of DNA transcription (Section 3.2.4), they cannot really be considered an example of gene control since virtually the whole of the genome (with the exception of some repeated DNA) participates in this amplification and there is no evidence for selective amplification of the genes for proteins required in the salivary gland.

DNA amplification and deletion in ciliated protozoa

A similar polytenization also occurs in many genera of ciliated Protozoa,

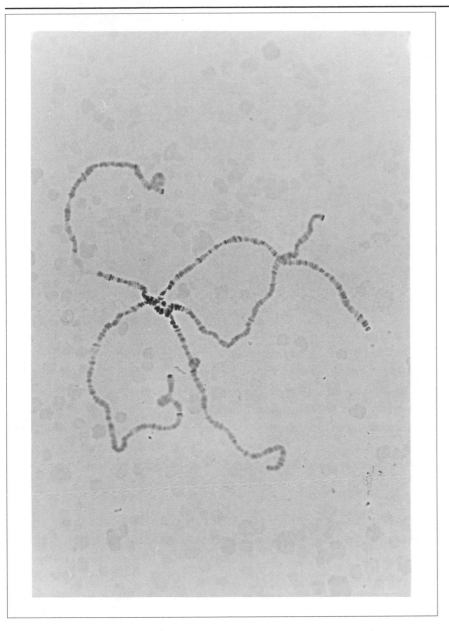

Figure 2.10 Polytene chromosomes of *Drosophila melanogaster.*

such as *Oxytrichia* and *Tetrahymena*. In these unicellular organisms development is controlled by separate germ line micronuclei and somatic macronuclei. As with the multicellular organisms discussed in Section 2.2.2, a selective loss of DNA occurs in the somatic nucleus, involving DNA required only in the germ line. Unlike in the multicellular organisms, however, this is preceded by a polytenization of the macronuclear DNA, which is followed by the fragmentation and degradation of up to 95% of the DNA (for review, see Prescott, 1992). Interestingly, the

Figure 2.11 Stages in the amplification of the macronuclear DNA of ciliated Protozoa.

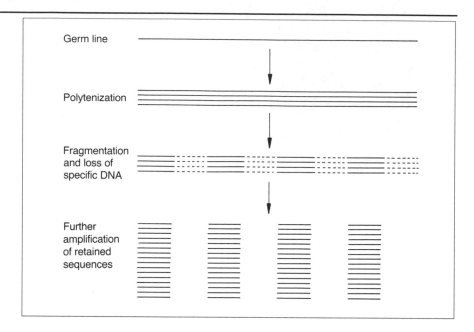

fragments that survive this degradation process and contain the genes required for somatic growth then undergo further rounds of amplification to produce the mature macronuclear DNA (Figure 2.11). Hence the somatic macronuclear DNA is created by a non-selective amplification followed by selective degradation and further amplification of the DNA. Presumably the many copies of the genes for structural proteins retained in the macronucleus are required to produce the large amounts of these proteins necessary in these very large single-celled organisms.

Amplification of ribosomal DNA
In addition to the amplification of chromosomal DNA in these ciliated Protozoa, a separate amplification event specifically increases the copy number of the DNA encoding the ribosomal RNA molecules. After fertilization, the single copy of this DNA present in the germ line is excised from the chromosomal DNA as a single 10.5 kb molecule and, after forming an almost perfect dimeric molecule, is subsequently extensively replicated extra chromosomally (Figure 2.12; Yao *et al.*, 1985).

Such amplification of the ribosomal DNA in order to produce the extensive amounts of structural RNAs required by the ribosomes is also observed in the embryonic development of multicellular organisms, even though the germ line DNA of these organisms contains multiple copies of the ribosomal DNA prior to any amplification event. Thus, early in oogenesis the frog oocyte, like all other frog cells, contains approximately 900 chromosomal copies of the DNA encoding ribosomal RNA. During

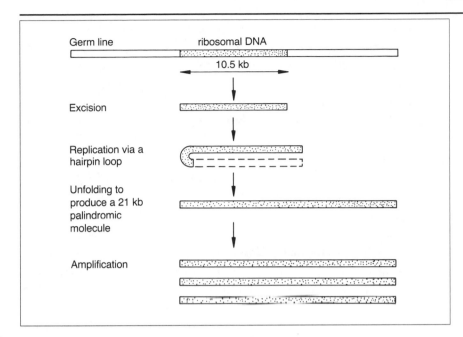

Figure 2.12 Stages in the amplification of the single copy of the DNA encoding the ribosomal RNA in ciliated Protozoa.

oogenesis, some of the copies are excised from the genome and replicate extensively to produce up to 2 million extra chromosomal copies of the ribosomal DNA (Brown and Dawid, 1968).

These cases of ribosomal DNA amplification clearly represent a response to the intense demand for ribosomal RNA during embryonic development. This requirement has to be met by DNA amplification since the ribosomal RNA represents the final product of ribosomal DNA expression. By contrast, in the case of protein-coding genes, high-level expression can be achieved both by very efficient transcription of the DNA into RNA and also by similarly efficient translation of the RNA into protein. Hence the observed amplification of ribosomal RNA genes does not imply that this mechanism is used in higher organisms to amplify specific protein-coding genes. Indeed the general similarity of the chromosome complement in different tissues argues against large-scale amplification of such genes. As with gene deletion, however, the sensitivity of such techniques is such that small-scale amplification of a few genes would not be detected and it is necessary to use molecular techniques to see whether such amplification occurs.

2.3.3 Molecular studies

In addition to its use in searching for gene loss (Section 2.2.4) the technique of Southern blotting can also be used to search for DNA amplifications in tissues expressing a particular gene. Thus, either new DNA bands hybridizing to the specific probe will appear, or the same band, present in other tissues, will be observed but it will hybridize more intensely because more copies are present. Despite very many studies of this type only a very small number of cases of such amplification have been reported, the vast majority of tissue-specific genes being present in the same number of copies in all tissues. Thus, for example, the fibroin gene is detectable at similar copy number in the DNA of the posterior silk gland of the silk moth, where it is expressed, and in the DNA of the middle silk gland, where no expression is detectable (Figure 2.13; Manning and Gage, 1978).

However, one exception to this general lack of amplification of protein-coding genes is particularly noteworthy. This involves the genes encoding the eggshell, or chorion, proteins in *Drosophila melanogaster*. As first shown by Spradling and Mahowald (1980), the chorion genes are selectively amplified (up to 64 times) in the DNA of the cells which surround the egg follicle, allowing the synthesis in these cells of the large amounts of chorion mRNA and protein needed to construct the eggshell. Such amplification can be observed readily in a Southern blot experiment using a recombinant DNA probe derived from one of the chorion genes (Figure 2.14). Unlike ribosomal DNA, amplification occurs within the chromosome without excision of the template DNA or the newly replicated copies.

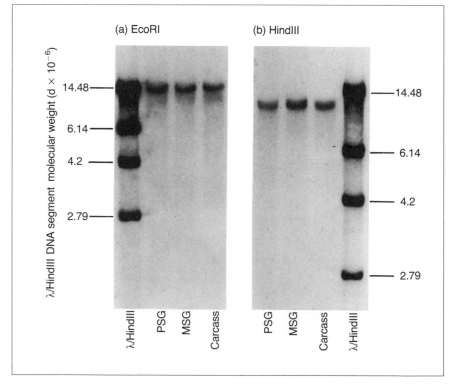

Figure 2.13 Southern blot of DNA prepared from the posterior silk gland (PSG), the middle silk gland (MSG) or the carcass of the silk moth, *Bombyx mori*, with a probe specific for the fibroin gene. Note the identical size and intensity of the band produced by *Eco*RI or *Hind*III digestion in each tissue, although the gene is only expressed in the posterior silk gland.

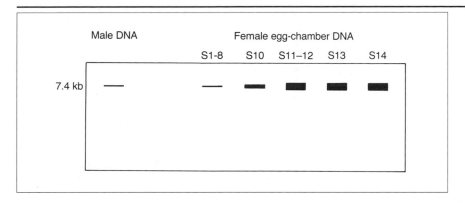

Figure 2.14 Amplification of the chorion genes in the ovarian follicle cells of *Drosophila melanogaster*. Note the dramatic increase in the intensity of hybridization of the chorion gene-specific probe to the DNA samples prepared from stage 10 to stage 14 egg chambers compared to that seen in stages 1–8 or in male DNA.

This amplification event is of particular interest because it occurs in normal cells as part of normal embryonic development, in contrast, for example, to the amplification of cellular oncogenes that occurs in some cancer cells (Section 8.3) or the amplification of the gene encoding the enzyme dihydrofolate reductase (which occurs in mammalian cells in response to exposure to the drug methotrexate; Schimke, 1980). None the less, the amplification of the chorion genes appears to be a response to a very specialized set of circumstances, requiring a novel means of gene regulation. Thus eggshell construction in *Drosophila* occurs over a very short period (about 5 h) and probably necessitates the synthesis of mRNA for the chorion proteins at very high rates, too large to be achieved even by high-level transcription of a single unique gene. It is this combination of high-level synthesis and a very short period to achieve it which necessitates the use of gene amplification. The restricted use that is made of this mechanism is well illustrated, however, by the fact that in silk moths, in which eggshell production occurs by a similar mechanism, multiple copies of the genes for the chorion proteins are present in the germ line. Thus they are present in all somatic cells, including those that do not express these genes, and are not further amplified in the ovarian follicle cells.

Such a finding serves to reinforce the conclusion of both molecular and chromosomal experiments that, as with DNA loss, DNA amplification is not a general method of gene regulation and is used only in isolated specialized cases.

2.4 DNA rearrangement

Having established that, generally, gene regulation does not occur by DNA deletion or amplification, it is necessary to consider the possibility

Figure 2.15 Model for activation of a gene by a DNA rearrangement involving either the removal of a repressor element (A) or the joining of the gene and its promoter (B).

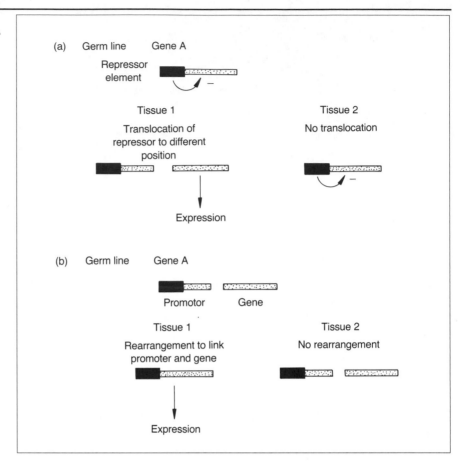

that it might occur via specific rearrangements of genes in tissues where their protein products are required. Such activation could occur, for example, by translocation of the gene, removing it from a repressive effect of neighboring sequences or by bringing together the part of the gene encoding the protein and a promoter element necessary for its transcription (Figure 2.15).

As with DNA amplification, the occurrence of such rearrangements in one particular cell type might not preclude the subsequent rearrangement of genes required in other tissues. Hence the ability of a single differentiated cell type to give rise to a variety of other differentiated cells in functional studies (Section 2.2.3) cannot be used to argue against the occurrence of such rearrangements. Similarly, these rearrangements might be on too small a scale to be detectable in chromosomal studies.

The evidence from molecular studies, however, shows unequivocally that these types of rearrangement are not involved in the selective activa-

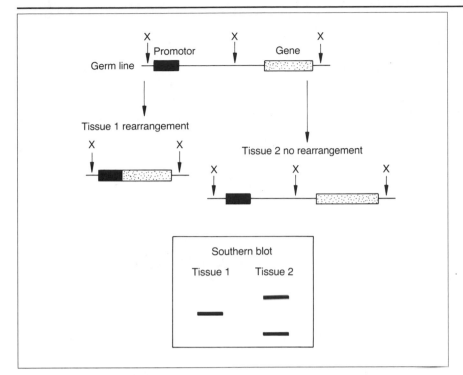

Figure 2.16 Generation of new restriction fragments detectable on a Southern blot following gene rearrangement. Sites for a particular restriction enzyme are indicated by an X and the consequences of digestion with this enzyme of unrearranged or rearranged DNA are indicated.

tion of the vast majority of genes in particular tissues. Thus such rearrangements would produce novel bands when the DNA from a tissue in which the gene was active was cut with restriction enzymes and used in Southern blot experiments (Figure 2.16). In fact no such novel bands are observed in the vast majority of cases when the DNA of a tissue in which the gene is active is compared with tissues where it is inactive (Figures 2.8 and 2.13).

The evidence against DNA rearrangement as a general means of gene control, obtained in such Southern blot experiments, has been abundantly confirmed by experiments in which the DNA sequence of a particular gene has been obtained both from a tissue where it is inactive and from one where it is active. Such experiments have failed to reveal even a single base difference in these two situations and indicate that the gene is potentially fully functional in tissues where it is inactive. Despite these clear findings that, as with deletion and amplification, DNA rearrangement is not a general means of gene control, a few cases where it is used have been reported and these will now be considered.

Yeast mating type
In yeast, mating involves the fusion of two opposite mating types, known as a and α, to generate a diploid cell. In some yeast strains, known as

Figure 2.17 Mating type switching in yeast. In every generation the larger mother cell, which has produced a smaller daughter, switches its genotype from *a* to *α* or vice versa.

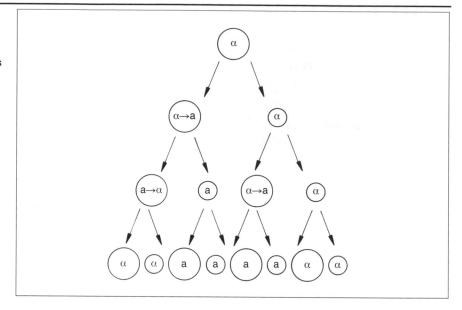

heterothallic strains, these two mating types are entirely separate, as in higher organisms. In homothallic yeasts, by contrast, a cell can switch its mating type from a to α or vice versa, and after such a switch it behaves exactly as does any other cell with its new mating type (reviewed by Nasmyth and Shore, 1987; Herskowitz, 1989). This switching takes place at a precise time in the life cycle, after a cell has given rise to a daughter by budding (Figure 2.17). Only the mother cell that produced the daughter undergoes switching; the daughter cell does not switch until it has grown and produced a daughter itself. Repeated cycles of switching from a to α and back occur in this way throughout the life of the organism. Although the reasons for this switching process are unclear, it probably represents a response to the need for rapid fusion of the haploid a and α cells to produce a diploid zygote. The switching of mating type in some of the progeny of a single cell allows this to occur without the requirement for contact with another strain of a different mating type.

A series of elegant genetic experiments showed that whether a cell was a or α in mating type was controlled by whether it possessed an *a* or *α* gene at the transcriptionally active *MAT* (or mating type locus) on chromosome 3. In addition to this active copy, yeast cells also possess transcriptionally silent copies of both *a* and α genes elsewhere on chromosome 3 (known as the *HML* and *HMR* loci). Switching occurs by a cassette mechanism in which one of these inactive copies replaces the active copy in the *MAT* locus. When an a gene in the *MAT* locus is replaced by an α gene, the α gene becomes active and the mating type of the cell switches (Figure 2.18).

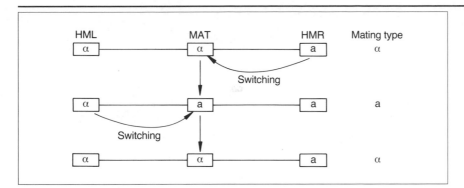

Figure 2.18 Mechanism of mating type switching involving the movement of an *a* or *α* gene from the inactive *HML* or *HMR* loci to the active *MAT* locus.

The reverse switch, in which an α gene at the *MAT* locus is replaced by an *a* gene, allows the switch from α to a to occur.

The process of switching is catalyzed by an endonuclease, which is the product of the *HO* (or homothallism) gene and which makes a double-stranded cut in the DNA at the *MAT* locus, initiating the transfer process. HO is only synthesized during a short part of the G_1 phase of the cell cycle. This is because the transcription of the HO gene is dependent on two regulatory proteins Swi4 and Swi6 which are only present in an active form during G_1 (for review, see Andrews and Mason, 1993). This finding does not explain, however, why transcription of HO should only occur in the G_1 phase in mother cells and not in daughter cells, leading to switching only in mother cells and not daughter cells.

In fact a further regulatory process restricts the transcription of the HO gene so that it occurs in the G_1 phase of the cell cycle only in mother cells. This regulation of *HO* transcription has been shown to involve a negative regulator known as SIN3, which represses *HO* transcription, and another regulator, Swi5, which allows transcription even in the presence of SIN3 (reviewed by Hicks, 1987). It has been postulated (Sternberg *et al.*, 1987) that both Swi5 and SIN3 are expressed at defined times in G_1, with Swi5 expressed first. In mother cells Swi5 and SIN3 are present together at G_1, the only point in the cell cycle when Swi4 and 6 are present and *HO* can therefore be expressed. Thus Swi5 antagonizes the repressive action of SIN3, and HO is made (Figure 2.19). In daughter cells, which are smaller, much more time is spent in G_1 and Swi5 levels have decayed away before the window in which *HO* can be expressed is reached. Hence when the cells reach G_1 where Swi4 and 6 are present, Swi5 is absent and only SIN3 is present, resulting in no transcription occurring.

Hence the regulation of HO transcription involves the interaction of two sets of regulatory elements, one of which restricts transcription to the G_1 phase of the cell cycle and the other which restricts it to mother and

Figure 2.19 Model for the regulation of *HO* gene transcription in mother and daughter cells by the antagonistic action of the SIN3 and Swi5 regulators. In mother cells (a) Swi5 expression overlaps that of SIN3 and antagonizes it, allowing HO to be made. In daughter cells (b), which are smaller and hence have a longer G$_1$ period, only SIN3 is present at the critical period for *HO* expression, SIN3 thus represses HO synthesis and switching does not occur.

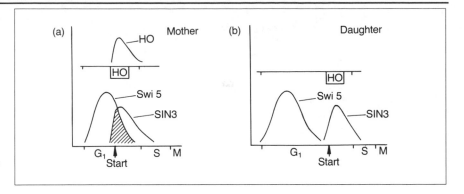

not daughter cells (Figure 2.20). The great attraction of this model is that it does not explain *HO* regulation merely in terms of an event, such as asymmetric partitioning of regulatory molecules between mother and daughter cells, which would in turn require an explanation. Rather, it relies on an entirely independent primary regulatory event, i.e. the difference in the timing of the cell division cycle between large and small cells. Thus it provides a mechanism for initiating the regulatory process rather than merely setting the question back one step.

This process, worked out by the use of the powerful genetic techniques available in yeast, may serve as a model for the regulation of developmental processes in higher organisms. Indeed, Herskowitz (1989) has drawn the analogy between the switching of mating type in yeast and the creation of a stem cell lineage in the development of higher organisms. Thus, if the diagram of switching shown in Figure 2.17 is modified, by assuming first that switching is irreversible and secondly by considering that α represents a stem cell while a represents a differentiated derivative, the model in Figure 2.21 is obtained. In this model lineage, a stem cell is continually

Figure 2.20 Summary of the regulatory mechanisms acting on the HO gene. The gene is transcribed only when Swi4 and 6 are present together with Swi5 which overcomes the inhibitory action of SIN3. The gene is thus transcribed only in the G$_1$ phase of the cell cycle in mother and not daughter cells.

Cell cycle	Mother/ daughter	Swi 5	Swi 4/6	Transcription
G1	M	+	+	+
G1	D	−	+	−
S/G2	M	−	−	−
S/G2	D	−	−	−

Sin 3   Swi 5   Swi 4   Swi 6
                −   +   +
→ HO gene

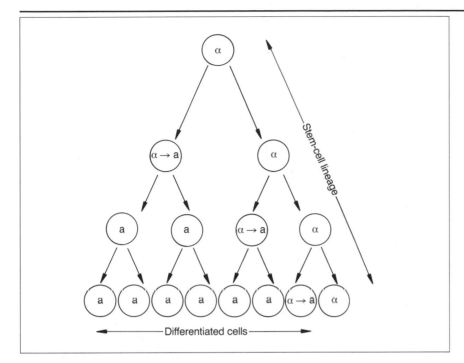

Figure 2.21 Model for the generation of a stem cell lineage, producing differentiated cells by a system based on α to a mating type switching. In this model, α represents the stem cell and a the differentiated cell, and it is assumed that, unlike in the yeast mating type system, the α to a switch is irreversible.

dividing, generating one daughter which differentiates while the other maintains the stem cell lineage. This is a very common feature in the development of higher organisms and has already been discussed in the erythrocyte lineage earlier in this chapter (Section 2.2.2).

Even if this model for the generation of a stem cell lineage is correct, it does not imply that the switch process need take place via a DNA rearrangement rather than by any other control mechanism that would allow the cell to change from one state to another. However, the possibility that such switches do occur in the development of higher organisms, by whatever means, needs to be considered and the mating type system offers a model for how this might be regulated.

Interestingly, the mating type system also involves other mechanisms of gene regulation that may be relevant to higher organisms. Thus the silent copies of the *a* and α genes in the HML and HMR loci are maintained in a transcriptionally inactive form within a tightly packed chromatin structure, a process which is also involved in the regulation of gene expression in mammals and other higher eukaryotes (Section 5.4). Similarly, the a and α gene products produce, respectively, the a and α phenotypes by activating or repressing the transcription of sets of other genes (the a- and α-specific genes). Most interestingly, the structure of the a and α gene products bears similarities to the homeobox proteins, which are regulatory proteins that regulate the transcription of other genes during the development of higher organisms such as mammals and *Drosophila* (Section 7.2.2). Hence the a and α proteins are members of a class of regulatory proteins which act in different organisms to control the transcription of other genes whose protein products are required for specific cellular processes.

It is clear, therefore, that the study of the yeast mating type system, which is very amenable to genetic analysis, can offer insights into the mechanisms being used to regulate gene expression in higher organisms. It is unlikely, however, that the use of DNA rearrangement to regulate the expression of the *a* and α control loci in this system indicates that this type of mechanism is used widely to control the expression of regulatory proteins in higher organisms. Rather, it appears to represent a response to the need to express *a* or α but not both, and to continually switch between the a and α states.

Antigenic variation in trypanosomes

Clearly, a similar switch mechanism to that used in yeast would be ideal in a situation where it was necessary to express successively one of a series of particular genes. In this situation one of many different genes present in inactive reservoir sites could be brought to the active site and expressed, only to be subsequently replaced by another gene from the reservoir. This mechanism is indeed used for precisely this purpose to control antigenic variation in trypanosomes (reviewed by van der Ploeg *et al.*, 1992). These unicellular Protozoa avoid the defensive response mounted by the immune system of the mammals they parasitize by continually changing the structure of their outer surface glycoprotein, which is accordingly known as the variant surface glycoprotein (VSG). This is achieved by repeatedly replacing the copy of the VSG gene at an expression site, with one of approximately 1000 different VSG genes present at other sites in the genome (Figure 2.22). At the expression site the VSG gene becomes linked to sequences required for its transcription and hence is actively transcribed. The copies present at other sites lack these sequences and hence they are inactive. The differences in the protein encoded by the different VSG genes that are successively placed in the active site thus produce the observed variation in the surface glycoprotein that occurs during the progress of an infection and prevents the mounting of an effective immune response by the parasitized host.

Antibody production in mammalian B cells

When mammals are exposed to foreign bacteria or viruses, their immune systems respond by synthesizing specific antibodies directed against the

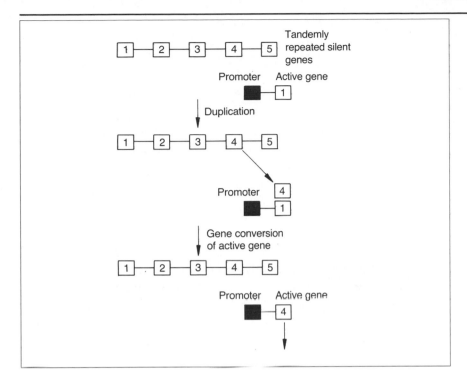

Figure 2.22 Switching of surface antigen expression in trypanosomes is achieved by the duplication of one of a tandem array of silent genes and its replacement of the active gene next to the promoter by a process of gene conversion.

proteins of these organisms, with the aim of neutralizing the harmful effects of the infection. The potential requirement for the synthesis of diverse kinds of antibody by the mammalian immune system far exceeds the diversity of different VSG proteins encoded by the trypanosome genome discussed in the preceding section. Thus, not only must the body be capable of producing different highly specific antibodies against all the different VSG proteins, but it must also be able to defend itself against a bewildering variety of challenges from other infectious organisms by the production of similarly specific antibodies against their proteins (for a recent review of the immune system, emphasizing the molecular aspects, see Roitt *et al.*, 1993). These antibodies, or immunoglobulins, are produced by the covalent association of two identical heavy chains and two identical light chains to produce a functional molecule (Figure 2.23). The combination of specific heavy and light chains produces the specificity of the antibody molecule. In particular, each chain contains, in addition to a constant region (which is relatively similar in different antibody molecules), a highly variable region which differs widely in amino acid sequence in different antibodies. It is this variable region that actually interacts with the antigen and hence determines the specificity of the antibody molecule.

As any heavy chain can associate with any light chain and as approxi-

Figure 2.23 Association of two heavy (H) and two light (L) chains to form a functional antibody molecule. The disulfide bonds linking the chains and the region that binds antigen are indicated.

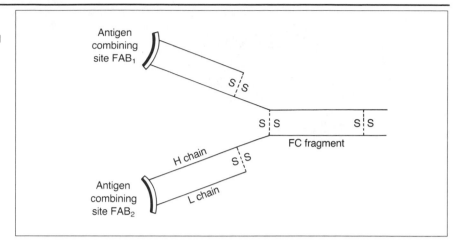

Figure 2.23 Association of two heavy (H) and two light (L) chains to form a functional antibody molecule. The disulfide bonds linking the chains and the region that binds antigen are indicated.

mately 1 million types of antibody specificities can be produced, at least 1000 genes encoding different heavy chains and a similar number encoding the light chains would be required. Copy-number studies have shown, however, that the germ line does not contain anything like this number of intact immunoglobulin genes and no specific amplification events have been detected in the DNA of antibody-producing B cells.

Rather, as first suggested by Dreyer and Bennett (1965), functional immunoglobulin genes are created by DNA rearrangements in the B cell lineage (reviewed by Tonegawa, 1983; Gellert, 1992). Thus the germ line DNA contains a large number of tandemly repeated DNA segments encoding different variable regions, which are separated by over 100 kbp from a much smaller number of DNA segments encoding the constant region of the molecule (Figure 2.24). This organization is maintained in most somatic cell types, but in each individual B cell a unique DNA rearrangement event occurs by which one specific variable region is brought together with one constant region by deletion of the intervening DNA. In this manner, a different functional gene, containing specific variable and constant regions, is produced in each individual B cell.

These rearrangements can be detected readily by the appearance of novel bands in Southern blot analyses of the structure of the immunoglobulin genes in the DNA of antibody-producing B cells. In the example shown in Figure 2.25 (based on the data of Hozumi and Tonegawa, 1976) the digestion of DNA from non-B cells with the restriction enzyme *Bam*HI produces fragments of 6.5 and 9.5 kb containing the variable and constant regions, respectively; whereas digestion of rearranged B cell DNA generates a single 3.5 kb band containing both parts of the gene. This combination of different variable (V) and constant (C) regions in the production of a functional immunoglobulin gene obviously creates a

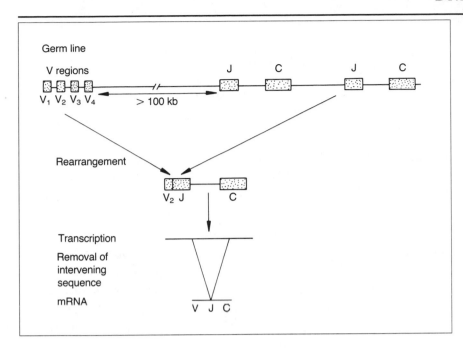

Figure 2.24 Rearrangement of the gene encoding the λ light chain of immuno-globulin results in the linkage of one specific variable region (V) to a specific joining region (J) with removal of the intervening DNA. The gene is then transcribed into RNA and the region between the joining (J) and constant regions (C) is removed by RNA splicing to create a functional mRNA.

wide range of different potential antibody specificities by the joining of particular V and C segments.

Despite this, more and more complex rearrangements are used in different types of immunoglobulin gene to generate still more potential variety. Thus, in the simplest case that we have discussed so far, that of the

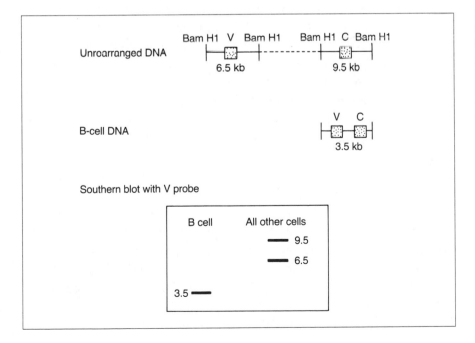

Figure 2.25 The rearrange-ment of the immunoglobulin gene locus creates a different *Bam*HI restriction fragment hybridizing to immunoglobu-lin DNA-specific probes in the DNA of B cells, compared to those that are present in other cell types where no rearrangement has occurred.

Figure 2.26 Rearrangement in the gene encoding the heavy chain of the immunoglobulin molecule involves joining of a variable (V) and a diversity (D) region as well as the similar joining of a diversity and a junction (J) region, with removal of the intervening DNA in each case.

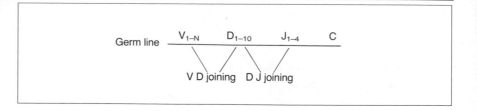

λ light chain (Figure 2.24), one of many V regions is brought into association with a constant region, consisting of a single J segment (encoding the junction between the variable and constant regions) which is separated by a short region from the DNA encoding the C segment itself. The rearranged gene is then transcribed and the intervening sequence between the J and C segments removed by RNA splicing to produce a functional mRNA (Section 4.2). In the genes for another type of light chain, known as κ, the situation is more complex. Thus these genes contain multiple J segments linked to individual constant segments and hence further diversity is created by the recombination of one of many V regions with one of several different J regions.

The case of the genes encoding the heavy-chain molecule is still more complex (Figure 2.26) in that these genes also contain a D, or diversity, segment (Early *et al.*, 1980). A functional heavy-chain gene is thus created by two successive rearrangements. In the first of these one of several D segments is brought together with one of the J segments and the associated C region. Subsequently, a V region segment is recombined with this DJC element exactly as occurs in the VJ joining found in the light-chain genes. These two combinatorial events allow the selection of one particular copy of each of four regions to produce the functional gene, generating many different possible heavy-chain molecules.

Interestingly, the generation of diversity by DNA rearrangement is supplemented by another DNA alteration, so far unique in mammalian cells, to the genes of the immune system. It appears that immunoglobulin genes can undergo point mutations within particular B cells, resulting in changes in the amino acids they encode and hence in their specificity. Such somatic mutation (reviewed by Thompson, 1992) produces forms of immunoglobulin not encoded by the DNA present in other tissues or in the germ line, and represents another form of tissue-specific gene alteration occurring within the immunoglobulin genes.

It is clear that the primary role of the rearrangements and alterations which occur in the immunoglobulin genes in B cells is the generation of the diversity of antibodies required to cope with the vast number of different possible antigens, rather than to produce the high-level expression of the immunoglobulin genes which occurs in such cells. None the less, high-level immunoglobulin gene expression does occur as a consequence of such rearrangements. Thus the DNA encoding the variable regions of the antibody molecule is closely linked to the promoter elements that direct the transcription of the gene. Such elements show no activity in the unrearranged DNA of non-B cells, but become fully active only when the variable region is joined to the constant region. In the case of the heavy-chain genes (Banerji *et al.*, 1983; Gilles *et al.*, 1983), this is because the intervening sequence between the J and C regions of these genes contains an enhancer element (Section 6.3) which, although not a promoter itself, greatly increases the activity of the promoter adjacent to the V region element. Hence the rearrangement not only brings the promoter element into a position where it can produce a functional mRNA containing VDJ and C regions but also facilitates the high-level transcription of this gene by juxtaposing promoter and enhancer elements that are separated by over 100 kb of DNA before the rearrangement event (Figure 2.27).

The element of gene control in this process is, however, entirely secondary to the need to produce diversity by rearrangement. Thus, if the immunoglobulin enhancer is linked artificially close to the immunoglobulin promoter and introduced into a variety of cells, it will activate transcription from the promoter only in B cells and not in other cell types. Hence the enhancer is active only in B cells and would not activate immunoglobulin transcription even if functional immunoglobulin genes containing closely linked VDJ and C segments were present in all other tissues. The immunoglobulin genes are therefore regulated by tissue-specific activator sequences in much the same way as other genes (Chapter 6), but such activation has been made to depend on the occurrence of a DNA rearrangement whose primary role lies elsewhere.

Thus, although similar DNA rearrangements are used in other situations where diversity must be generated, for example in the genes encoding

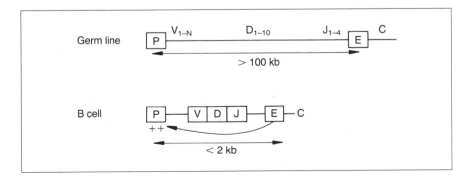

Figure 2.27 The rearrangement of the heavy-chain gene brings an enhancer (E) in the intervening sequence between J and C close to the promoter (P) adjacent to V, and results in the activation of the promoter and gene transcription.

the T cell receptor which is involved in the response to foreign antigen of T lymphocytes (reviewed by Davis, 1985), it is clear that they do not represent a general means of gene regulation applicable to other situations. Such a conclusion is entirely in accordance with the results of Southern blot and sequencing experiments which, as previously discussed, have failed to detect such rearrangements in the vast majority of genes active only in particular tissues.

2.5 Conclusions

Although we have discussed in this chapter a number of cases where DNA loss, amplification or rearrangement regulate gene expression, it is clear that these represent isolated special cases where the dictates of a particular situation have necessitated the use of such mechanisms. Thus the loss of DNA in red blood cells is dictated by the need to fill the cell with hemoglobin, the amplification of the chorion genes in *Drosophila* by the requirement to produce the corresponding proteins in a very short time and the rearrangement of the immunoglobulin genes by the need to generate a diverse array of antibodies. Hence these cases are not representative of a general mechanism of gene regulation by DNA alteration and, in general, the DNA of different cell types is quantitatively and qualitatively identical. Given that we have established previously that the RNA content of different cell types can vary dramatically, it is now necessary to investigate how such differences in RNA content can be produced from the similar DNA present in such cell types.

References

Andrews, B. J. and B. W. Mason 1993. Gene expression and the cell cycle: a family affair. *Science* **261**, 1543–44.

Baltimore, D. 1981. Somatic mutation gains its place among the generators of diversity. *Cell* **26**, 295–6.

Banerji, J., L. Olson and W. Schaffner 1983. A lymphocyte-specific cellular enhancer is located downstream of the joining region in immunoglobulin heavy chain genes. *Cell* **33**, 729–40.

Bessis, N. and M. Brickal 1952. Aspect dynamique des cellules du sang. *Revues Hématologique* **7**, 407–35.

Brown, D. D. 1981. Gene expression in eukaryotes. *Science* **211**, 667–74.

Brown, D. D. and B. Dawid 1968. Specific gene amplification in oocytes. *Science* **160**, 272–80.

Davis, M. 1985. Molecular genetics of the T cell receptor beta chain. *Annual Review of Immunology* **3**, 537–60.

Dreyer, W. J. and J. C. Bennett 1965. The molecular basis of antibody formation: a paradox. *Proceedings of the National Academy of Sciences of the USA* **54**, 864–9.

Early, P., H. Huang, M. Davis, K. Calame and L. Hood 1980. An immunoglobulin heavy chain variable region is generated from three segments of DNA: V_H, D and J_H. *Cell* **19**, 981–92.

Gellert, M. 1992. Molecular analysis of V(D)J recombination. *Annual Review of Genetics* **22**, 425–46.

Gillies, S. D., S. L. Morrison, V. T. Oi and S. Tonegawa 1983. A tissue specific enhancer is located downstream of the joining region in immunoglobulin heavy chain genes. *Cell* **33**, 717–28.

Gurdon, J. B. 1968. Transplanted nuclei and cell differentiation. *Scientific American* **219** (December), 24–35.

Herskowitz, L. 1989. A regulatory hierachy for cell specialization in yeast. *Nature* **342**, 749–57.

Hicks, J. B. 1987. Mechanisms of differentiation. *Nature* **326**, 444–5.

Hozumi, N. and S. Tonegawa 1976. Evidence for somatic rearrangement of immunoglobulin genes coding for variable and constant regions. *Proceedings of the National Academy of Sciences of the USA* **73**, 3628–32.

Jeffreys, A. J. and R. A. Flavell 1977. The rabbit β-globin gene contains a large insert in the coding sequence. *Cell* **12**, 1097–108.

John, B. and G. L. G. Miklos 1979. Functional aspects of satellite DNA and heterochromatin. *International Review of Cytology* **58**, 1–113.

Kafatos, F. C., W. Orr and C. Delidakis 1985. Developmentally regulated gene amplification. *Trends in Genetics* **1**, 301–6.

Manning, R. F. and L. P. Gage 1978. Physical map of the *Bombyx mori* DNA containing the gene for silk fibroin. *Journal of Biological Chemistry* **253**, 2044–52.

Nasmyth, K. A. and D. Shore 1987. Transcriptional regulation in the yeast life cycle. *Science* **237**, 1162–70.

Prescott, D. M. 1992. Cutting, splicing, reordering and elimination of DNA sequences in hypotrichous ciliates. *Bioessays* **14**, 317–24.

Roitt, I. M., J. Brostoff and D. K. Male 1993. *Immunology.* Mosby, St Louis, MO.

Schimke, R. T. 1980. Gene amplification and drug resistance. *Scientific American* **243** (November), 50–9.

Southern, E. M. 1975. Detection of specific sequences among DNA fragments separated by gel electrophoresis. *Journal of Molecular Biology* **98**, 503–17.

Spradling, A. C. and A. P. Mahowald 1980. Amplification of genes for chorion proteins during oogenesis in *Drosophila melanogaster*. *Proceedings of the National Academy of Sciences of the USA* **77**, 1096–100.

Stark, G. R. and G. M. Wahl 1984. Gene amplification. *Annual Review of Biochemistry* **53**, 447–91.

Sternberg, P. W., M. J. Stern, I. Clark and I. Herskowitz 1987. Activation of the yeast *HO* gene by release from multiple negative controls. *Cell* **48**, 567–77.

Steward, F. C. 1970. From cultured cells to whole plants: the induction and control of their growth and morphogenesis. *Proceedings of the Royal Society, Series B* **175**, 1–30.

Thompson, C. B. 1992. Creation of immunoglobulin diversity by intra-chromosomal gene conversion. *Trends in Genetics* **8**, 416–22.

Tonegawa, S. 1983. Somatic generation of antibody diversity. *Nature* **302**, 575–81.

van der Ploeg, L. H. T., K. Gotterdiner and M. G. S. Lee 1992. Antigenic variation in African Trypanosomes. *Trends in Genetics* **8**, 452–7.

Wilson, E. B. 1928. *The Cell in Development and Heredity*. Macmillan, New York.

Yamada, T. 1967. Cellular and sub-cellular events in Wolffian lens regeneration. *Current Topics in Developmental Biology* **2**, 249–83.

Yao, M.-C., S.-G. Zhu and C.-H. Yao 1985. Gene amplification in *Tetrahymena thermophila*: formation of extra-chromosomal palindromic genes coding for rRNA. *Molecular and Cellular Biology* **5**, 1260–7.

Regulation at transcription

3

3.1 Levels of gene regulation

The observation that differences in the RNA and protein content of different tissues are not paralleled by significant differences in their DNA content indicates that the process whereby DNA produces mRNA must be the level at which gene expression is regulated in eukaryotes.

In bacteria this process involves only a single stage, that of transcription, in which an RNA copy of the DNA is produced by the enzyme RNA polymerase. Even while this process is still occurring, ribosomes attach to the nascent RNA chain and begin to translate it into protein. Hence cases of gene regulation in bacteria, such as the switching on of the synthesis of the enzyme β-galactosidase in response to the presence of lactose (its substrate), are mediated by increased transcription of the appropriate gene (Jacob and Monod, 1961). Clearly, a similar regulation of gene transcription in different tissues, or in response to substances such as steroid hormones which induce the synthesis of new proteins, represents an attractive method of gene regulation in eukaryotes. In contrast to the situation in bacteria, however, a number of stages intervene between the initial synthesis of the primary RNA transcript and the eventual production of mRNA (for review, see Nevins, 1983) (Figure 3.1). The initial transcript is modified at its 5' end, by the addition of a cap structure containing a modified guanosine residue, and is subsequently cleaved near its 3' end, followed by the addition of up to 200 adenosine residues in a process known as polyadenylation. Subsequently, intervening sequences, or introns, which interrupt the protein-coding sequence in both the DNA and the primary transcript of many genes (Jeffreys and Flavell, 1977; Tilghman *et al.*, 1978) are removed by a process of RNA splicing (for reviews, see Sharp, 1987; Lamond, 1991). Although this produces a functional mRNA, the spliced molecule must then be transported from the nucleus, where these processes occur, to the cytoplasm where it can be translated into protein.

Figure 3.1 Stages in eukaryotic gene expression which could be regulated.

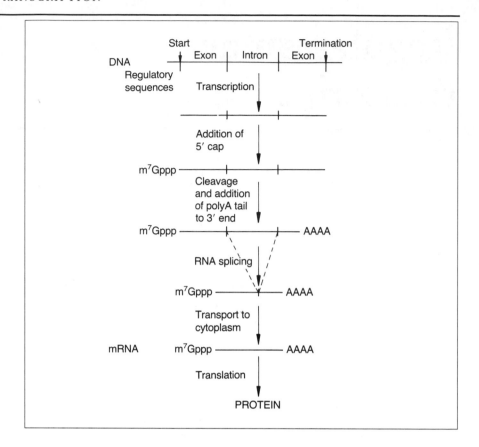

In theory any one of these stages, all of which are essential for the production of a functional mRNA, could be used to regulate the expression of specific genes in particular tissues. For example, it has been suggested that gene regulation could occur by transcribing all genes in all tissues and selecting which transcripts were appropriately spliced by correct removal of intervening sequences, thereby producing a functional mRNA (Davidson and Britten, 1979). The existence of such a plethora of possible regulatory stages has led, therefore, to much investigation as to whether all or any of these are used. In general, such studies have shown that in higher eukaryotes, as in bacteria, the primary control of gene expression is at the level of transcription, and the evidence showing that this is the case is discussed in this chapter. A number of cases of post-transcriptional regulation do exist, however, and these are discussed in Chapter 4.

3.2 Evidence for transcriptional regulation

The evidence for the regulation of gene transcription comes from several types of study, which will be considered in turn.

3.2.1 Evidence from studies of nuclear RNA

If regulation of gene expression takes place at the level of transcription, the differences in the cytoplasmic levels of particular mRNAs which occur between different tissues should be paralleled by similar differences in the levels of these RNAs within the nuclei of different tissues. In contrast, regulatory processes in which a gene was transcribed in all tissues and the resulting transcript either spliced or transported to the cytoplasm in a minority of tissues would result in cases where differences in mRNA content occurred without any corresponding difference in the nuclear RNA (Figure 3.2). Hence a study of the level of particular RNA species in the nuclear RNA of individual tissues or cell types serves as an initial test to distinguish transcriptional and post-transcriptional regulation.

The earliest studies in this area focused on the highly abundant RNA species produced during terminal differentiation, which could be readily studied simply because of their abundance. Thus even before the discovery of intervening sequences revealed the possibility of regulation at the level of RNA splicing, Gilmour *et al.* (1974) studied the processes regulating the accumulation of globin which occurs when Friend erythroleukemia cells are treated with an inducer of globin production, dimethyl sulfoxide. In this system, the accumulation of cytoplasmic globin RNA, which produces the increase in synthesis of globin protein, was paralleled by increasing accumulation of globin RNA within the nucleus, suggesting that the primary effect of the inducer was at the level of globin gene transcription. Similarly, in the experiments of Groudine *et al.* (1974), globin-specific RNA was readily detectable in the nuclei of globin-synthesizing

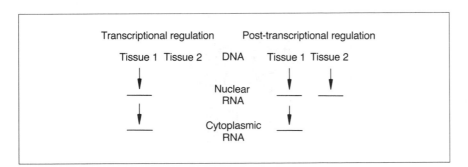

Figure 3.2 Consequences of transcriptional or post-transcriptional regulation on the level of a specific nuclear RNA in a tissue that expresses the corresponding cytoplasmic mRNA (tissue 1) and one that does not (tissue 2).

erythroblasts but not in the nuclear RNA of fibroblasts or muscle cells, hence indicating the involvement of transcriptional control in regulating the abundant production of globin RNA only in the red blood cell lineage and not in other cell types.

Subsequently, the use of Northern blotting (Section 1.3.2) allowed the separation of different species within nuclear RNA by size and their visualization by hybridization to an appropriate probe. In the case of genes with many intervening sequences, such as that encoding the egg protein ovalbumin, many different RNA species could be observed in the nucleus, including not only the primary transcript and the fully spliced RNA prior to transport to the cytoplasm but also a series of intermediate-sized RNAs from which some of the intervening sequences had still to be removed (Roop *et al.*, 1978; Figure 3.3). The identification of such potential precursors of the mature mRNA allowed a study of their expression in tissues either producing or not producing ovalbumin mRNA and protein. Thus cytoplasmic ovalbumin mRNA and protein are present only in the oviduct following stimulation with estrogen and disappear when estrogen

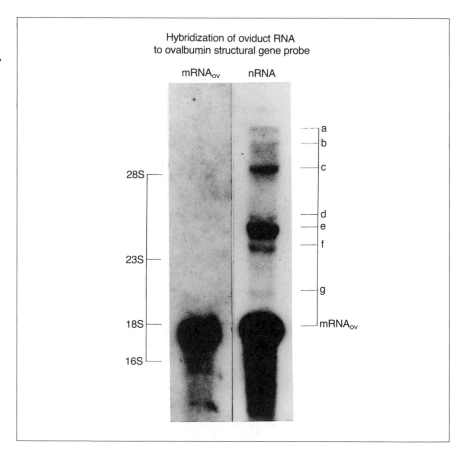

Figure 3.3 Northern blot showing that unspliced and partially spliced precursors (a–g) to the ovalbumin mRNA (mRNA$_{ov}$) are detectable in the nuclear RNA (nRNA) of estrogen-stimulated oviduct tissue.

is withdrawn. Similarly, the mRNA and protein are absent in other tissues, such as the liver, and cannot be induced by treatment with the hormone in these tissues. Studies of the distribution of both the fully spliced nuclear RNA and the larger precursors (Roop *et al.*, 1978) showed that these species could only be detected in the nuclear RNA of the oviduct following estrogen stimulation and were absent in the liver nuclear RNA, or in oviduct nuclear RNA following estrogen withdrawal (Figure 3.4). Hence the distribution of these precursors in the nucleus exactly parallels that of the cytoplasmic mRNA, a finding entirely consistent with the transcriptional induction of the ovalbumin gene in the oviduct in response to estrogen. These observations are difficult to reconcile, however, with a model in which the hormone acts to relieve a block in RNA splicing or transport, which exists in untreated oviduct and other tissues, since such models would predict an accumulation of unspliced or untransported RNA for ovalbumin within the nucleus.

These early studies have now been abundantly supplemented by many others measuring the nuclear RNA levels of other specific genes whose expression changes in particular situations, such as those encoding a number of highly abundant mRNAs present only in the soya bean embryo (Goldberg *et al.*, 1981) or that encoding the developmentally regulated mammalian liver protein, α-fetoprotein (Latchman *et al.*, 1984). In general, such studies have led to the conclusion that in most cases alterations

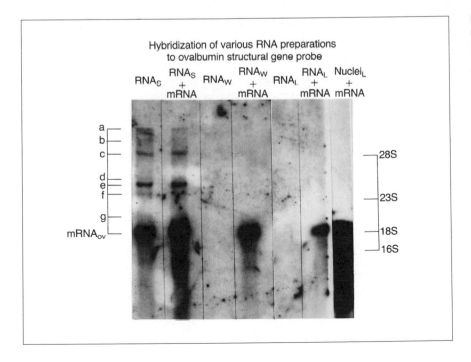

Figure 3.4 Northern blot showing that the nuclear precursors to ovalbumin mRNA (seen in Figure 3.3) are detectable in the nuclear RNA of estrogen-stimulated oviduct (RNA$_S$) but are absent in the nuclear RNA of estrogen-withdrawn oviduct (RNA$_W$) and of liver (RNA$_L$). Moreover, ovalbumin mRNA mixed with withdrawn oviduct (RNA$_W$ + mRNA) or liver (RNA$_L$ + mRNA) nuclear RNA is not degraded, showing that the absence of RNA for ovalbumin in these nuclear RNAs is not due to a nuclease specifically degrading the ovalbumin RNA.

in specific mRNA levels in the cytoplasm are accompanied by parallel changes in the levels of the corresponding nuclear RNA species.

Interestingly, such studies carried out with cloned DNA probes for individual RNA species of relatively high abundance are in contrast to reports using R_0t curve analysis (Section 1.3.1) to study variations in the total nuclear RNA population in different tissues. Such experiments, which examine mainly low-abundance RNAs, have suggested that in some organisms, such as sea urchins (Wold *et al.*, 1978) and tobacco plants (Kamaly and Goldberg, 1980), the nuclear RNA is a highly complex mixture of different RNAs, some of which can be retained in the nucleus in one tissue and transported to the cytoplasm in other tissues. These studies, which are indicative of post-transcriptional control in these organisms, are discussed in Chapter 4 (Section 4.1). It is noteworthy, however, that in mammals a different situation exists. Thus in these organisms the increased number of different sequences in nuclear compared with cytoplasmic RNA can be accounted for by the presence of intervening sequences, which are transcribed but not transported to the cytoplasm, without the need to postulate the existence of whole transcripts which are confined to the nucleus in specific tissues.

Hence it appears from these studies that, at least in mammals and other higher eukaryotes, the regulation of transcription, resulting in parallel changes in nuclear and cytoplasmic RNA levels, is the primary means of regulating gene expression. However, the studies described so far suffer from the defect that they measure only steady-state levels of specific nuclear RNAs. It could be argued that the gene is being transcribed in the non-expressing tissue but that the transcript is degraded within the nucleus at such a rate that it cannot be detected in assays of steady-state RNA levels (Figure 3.5). This degradation in the non-expressing tissue but not in the expressing tissue could be produced directly by specific regulation of the rate of turnover of a particular RNA in the different tissues, producing rapid degradation only in the non-expressing tissue.

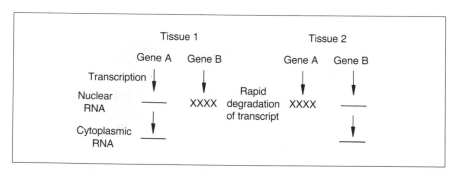

Figure 3.5 Model in which lack of expression of genes in particular tissues is caused by rapid degradation of their RNA transcripts.

Alternatively, it might result from a block to splicing or transport of the RNA in the non-expressing tissue, followed by the rapid degradation of this unspliced or untransported RNA, which in the expressing tissue would be processed or transported before it could be degraded. These considerations necessitate the direct measurement of gene transcription itself in the different tissues in order to establish unequivocally the existence of transcriptional regulation. Methods to do this have been devised and these will now be discussed.

3.2.2 Evidence from pulse-labeling studies

The synthesis of RNA from DNA by the enzyme RNA polymerase involves the incorporation of ribonucleotides into an RNA chain. Therefore the synthesis of any particular RNA can be measured by adding a radioactive ribonucleotide (usually uridine labeled with tritium) to the cells and measuring how much radioactivity is incorporated into RNA specific for the gene of interest. Clearly, if the degradation mechanisms discussed in the last section do exist, they will, given time, degrade the radioactive RNA molecule produced in this way and drastically reduce the amount of labeled RNA detected. In order to prevent this, the rate of transcription is measured by exposing cells briefly to the labeled uridine in a process referred to as pulse labeling. The labeled uridine is incorporated into nascent RNA chains that are being made at this time and, even before a complete transcript has had time to form, the cells are lysed and total RNA is isolated from them. This RNA, which contains labeled partial transcripts from genes active in the tissue used, is then hybridized to a piece of DNA derived from the gene of interest, the number of radioactive counts that bind providing a measure of the incorporation of labeled precursor into its corresponding RNA (Figure 3.6).

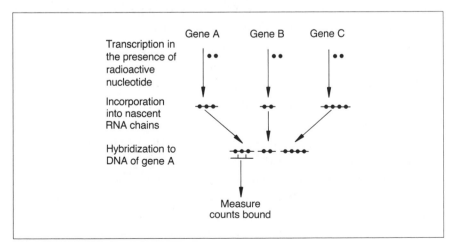

Figure 3.6 Pulse-labeling assay to assess the transcription rate of a specific gene (gene A) by measuring the amount of radioactivity (stars) incorporated into nascent transcripts.

This method provides the most direct means of measuring transcription and has been used, for example, to show that the induction of globin production which occurs in Friend erythroleukemia cells in response to treatment with dimethyl sulfoxide is mediated by increased transcription of the globin gene (Lowenhaupt *et al.*, 1978). This increased transcription produces the increased levels of globin-specific RNA in the nucleus and cytoplasm of the treated cells, which was discussed earlier in this chapter (Section 3.2.1).

Although pulse labeling provides a very direct measure of transcription rates, the requirement to use very short labeling times to minimize any effects of RNA degradation limits its applicability. Thus in the experiments of Lowenhaupt *et al.* (1978) it was possible to measure the amount of radioactivity incorporated into globin RNA in the very brief labeling times used (5–10 min) only because of the enormous abundance of globin RNA and the very high rate of transcription of the globin gene. With other RNA species, the rates of transcription are insufficient to provide measurable incorporation of label in the short pulse time. More label will, of course, be incorporated if longer pulse times are used, but such pulse times allow the possibility of RNA turnover and are therefore subject to the same objections as the measurement of stable RNA levels.

Hence, although pulse labeling can be used to establish unequivocally that transcriptional control is responsible for the massive synthesis of the highly abundant RNA species present in terminally differentiated cells, it cannot be used to demonstrate the generality of transcriptional control processes and, in particular, their applicability to RNAs which, although regulated in different tissues, never become highly abundant.

This limitation of the pulse-labeling method is especially relevant in view of the existence of the Davidson and Britten model for gene regulation (Davidson and Britten, 1979) which specifically postulates that highly abundant RNA species will be regulated in a different manner to the bulk of RNA species, which are of moderate or low abundance. Thus, in this model (Figure 3.7) all genes are postulated to be transcribed in all tissues at a low basal rate and regulation takes place at a post-transcriptional level by deciding which transcripts are processed to functional mRNA and which are degraded. In the case of most genes the level of RNA and protein required in any particular tissue would be met by processing correctly all of the primary transcript produced by this low basal rate of transcription. The level of the abundant RNA species would be such, however, that they could not be produced with this low rate of gene transcription, even by correctly processing all of the primary transcript. Hence, for the genes encoding these RNA species, a special

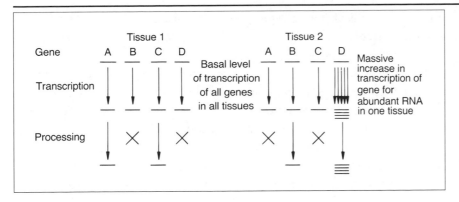

Figure 3.7 Davidson and Britten model of post-transcriptional regulation, in which all genes are transcribed at a low basal rate and regulation is achieved by controlling whether the resulting transcript is processed to mature mRNA. Transcriptional control is confined to genes (such as gene D) where the level of RNA required cannot be achieved by processing all the primary transcript produced by the low basal rate of transcription.

mechanism would operate and their transcription would be dramatically increased in some tissues, as observed by pulse labeling. This theory postulates that the regulation of gene expression by changes in transcription is confined to the few genes encoding highly abundant RNA species and that post-transcriptional control processes will regulate the expression of less abundant RNAs whose genes will be transcribed even in tissues where no mRNA is synthesized. In order to test this theory it is necessary to use a method of measuring transcription which, although less direct than pulse labeling, is more sensitive and hence can be applied to a wider variety of cases, including non highly abundant mRNAs. This method is discussed in the next section.

3.2.3 Evidence from nuclear run-on assays

The primary limitation on the sensitivity of pulse labeling is the existence within the cell of a large pool of non-radioactive ribonucleotides, which are normally used by the cell to synthesize RNA. When labeled ribonucleotide is added to the cell, it is considerably diluted in this pool of unlabeled precursor. The amount of label incorporated into RNA in the labeling period is therefore very small, since the majority of ribonucleotides incorporated are unlabeled. The sensitivity of this method is thus severely reduced, resulting in its observed applicability only to genes with very high rates of transcription. Interestingly, however, although transcription takes place in the nucleus, most of the pool of precursor ribonucleotides is present in the cytoplasm. Hence it is possible, by removing the cytoplasm and isolating nuclei, to remove much of the pool of unlabeled ribonucleotide. Labeled ribonucleotides can then be added directly to the isolated nuclei in a test tube and their rate of incorporation into RNA used as a measure of the rate of transcription. Because the labeled ribonucleotides are not diluted in the unlabeled cytoplasmic pool, considerably more label is incorporated into any particular gene than is observed in pulse-labeling experiments. The label incorporated into any particular transcript is detected by hybridization to its corresponding DNA exactly as in pulse-labeling experiments.

This method, which is known as a nuclear run-on assay, is therefore much more widely applicable than pulse labeling and can be used to quantify the transcription of genes that are never transcribed at levels detectable by pulse labeling. Moreover, very many studies (for review, see Darnell, 1982 and references therein) have now established that the RNA synthesized by isolated nuclei in the test tube is similar to that made by intact whole cells, and that the method is therefore not only sensitive but also provides an accurate measure of transcription, free from artefact.

In initial studies, nuclear run-on assays were used to measure the transcription of highly abundant RNA species. Thus, for example, nuclei isolated from the erythrocytes of adult chickens (which, unlike the equivalent cells in mammals, do not lose their nuclei) were shown to transcribe only the gene encoding the adult β-globin protein, whereas nuclei prepared from similar cells isolated from embryonic chickens failed to transcribe this gene and instead transcribed the gene encoding the form of β-globin made in the embryo (Groudine et al., 1981). Such a finding parallels the observation discussed earlier that the RNA specific for the adult form of β-globin is present only in adult and not in embryonic erythroid cell nuclei (Section 3.2.1) and indicates that the developmentally regulated production of different forms of globin protein is under transcriptional control.

A similar parallelism between the results of nuclear RNA studies and nuclear run-on assays of transcription is also found in the case of the ovalbumin gene. Thus the presence of ovalbumin-specific RNA only in the nuclear RNA of hormonally stimulated oviduct tissue is paralleled by the ability of nuclei prepared from stimulated oviduct cell nuclei to transcribe the ovalbumin gene at high levels in run-on assays. In contrast, nuclei from other tissues or unstimulated oviduct cells failed to transcribe this gene, paralleling the observed absence of ovalbumin RNA in the nuclei and cytoplasm of these tissues (Swaneck et al., 1979). Hence the observed increase in nuclear and cytoplasmic RNA for ovalbumin in response to estrogen is indeed caused by increased transcription of the ovalbumin gene. Interestingly, in this study the incorporation of label into ovalbumin RNA in the run-on assay was observed to peak after 15 min and did not decrease at longer labeling times of up to 1 h. This suggested that, unlike intact cells, isolated nuclei do not degrade or process the RNA that they synthesize and hence allowed the use of longer labeling times, further increasing the sensitivity of this technique.

This increased sensitivity has allowed the use of nuclear run-on assays to demonstrate that transcriptional control of many genes encoding specific proteins is responsible for the previously observed differences in the levels of these proteins in different situations. Cases where such transcriptional control has been demonstrated in this manner are far too numerous to mention individually but involve a range of different tissues and organisms, such as the synthesis of α-fetoprotein in fetal but not adult mammalian liver, production of insulin in the mammalian pancreas, the expression of the *Drosophila melanogaster* yolk protein genes only in ovarian follicle cells, the expression of aggregation stage-specific genes in the slime mould *Dictyostelium discoideum* and the expression of soya bean seed proteins such as glycinin in embryonic but not adult tissues.

These studies on individual genes for particular proteins have been supplemented by the more general studies of Darnell and colleagues (Derman *et al.*, 1981; Powell *et al.*, 1984). In these experiments the authors studied 12 different genes whose corresponding mRNAs were present in mouse liver cytoplasm but were absent in brain cytoplasm. These included both genes encoding previously isolated liver-specific proteins, such as albumin or transferrin, and those which had been isolated simply on the basis of the presence of their corresponding cytoplasmic RNA in the liver and not in other tissues and for which the protein product had not been identified. Measurement of the transcription rate of these genes in nuclei isolated from brain or liver showed that such transcription was detectable only in the liver nuclei (Figure 3.8), indicating that the difference in cytoplasmic mRNA level was produced by a corresponding difference in gene transcription. In these experiments the rate of

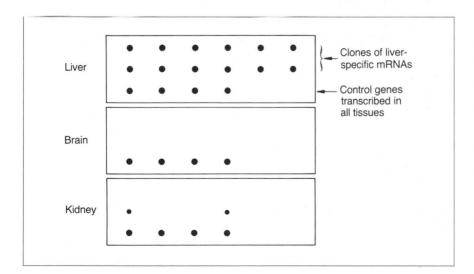

Figure 3.8 Nuclear run-on assay to measure the transcription rates of genes encoding liver-specific mRNAs and of control genes expressed in all tissues. Note that the liver-specific genes are not transcribed at all in brain, while in the kidney the only two of these genes to be transcribed are those that are known to produce a low level of mRNA in the kidney.

transcription of the 12 genes was also measured in nuclei prepared from kidney tissue. In this tissue mRNAs corresponding to two of the genes were present at a considerably lower level than that observed in the liver, while RNA corresponding to the other 10 genes was undetectable. As with the brain nuclei, the level of transcription detectable in the kidney nuclei exactly paralleled the level of RNA present. Thus, only the two genes producing cytoplasmic mRNA in the kidney were detectably transcribed in this tissue and the level of transcription of these genes was much lower than that seen in the liver nuclei.

These studies on a relatively large number of different liver RNAs of different abundances, when taken together with the studies of many genes encoding specific proteins, indicate that for genes expressed in one or a few cell types, increased levels of RNA and protein in a particular cell type are brought about primarily by increases in gene transcription and that, at least in mammals, post-transcriptional control mechanisms such as that postulated by Davidson and Britten (1979) are not the primary means of regulating gene expression, although they may be more important in other animals, such as the sea urchin (Section 4.1).

3.2.4 Evidence from polytene chromosomes

Although pulse-labeling and nuclear run-on studies of many genes have conclusively established the existence of transcriptional regulation, it is necessary to discuss another means of demonstrating such regulation, in which increased transcription can be directly visualized. As described in Chapter 2 (Section 2.3.2), the chromosomal DNA in the salivary glands of *Drosophila* is amplified many times, resulting in a giant polytene chromosome. Such chromosomes exhibit along their length areas known as puffs in which the DNA has decondensed into a more open state, resulting in the expansion of the chromosome (Figure 3.9). If cells are allowed to incorporate labeled ribonucleotides into RNA and the resulting RNA is then hybridized back to the polytene chromosomes, it localizes primarily to the positions of the puffs. Hence these puffs represent sites of intense transcriptional activity which, because of the large size of the polytene chromosomes, can be directly visualized. Most interestingly, many procedures which in *Drosophila* result in the production of new proteins, such as exposure to elevated temperature (heat shock) or treatment with the steroid hormone ecdysone, also result in the production of new puffs at specific sites on the polytene chromosomes, each treatment producing a different specific pattern of puffs. This suggests that these sites contain the genes encoding the proteins whose synthesis is increased by the treatment and that this increased synthesis is mediated via increased transcription of

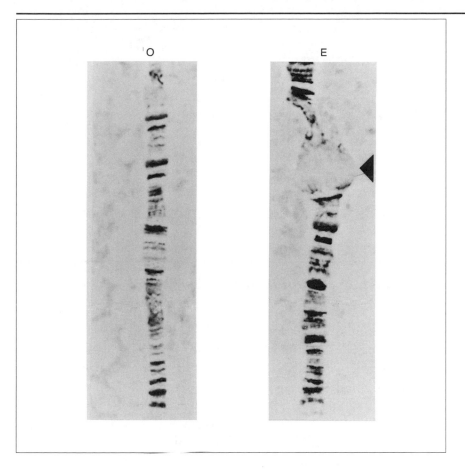

Figure 3.9 Transcriptionally active puff (arrowed) in a polytene chromosome of *Drosophila melanogaster*. The puff appears in response to treatment with the steroid hormone ecdysone (E) and is not present prior to hormone treatment (O).

these genes, which can be visualized in the puffs. In the case of ecdysone treatment this has been confirmed directly by showing that the radio-active RNA synthesized immediately after ecdysone treatment hybridizes strongly to the ecdysone-induced puffs but not to a puff which regresses upon hormone treatment. In contrast, RNA prepared from cells prior to ecdysone treatment hybridizes only to the hormone-repressed puff and not to the hormone-induced puffs (Figure 3.10; Bonner and Pardue, 1977). Similarly, RNA labeled after heat shock hybridizes intensely to puff 87C, which appears following exposure to elevated temperature and is now known to contain the gene encoding the 70 kDa heat-shock protein (hsp70) which is the major protein made in *Drosophila* following heat shock (Spradling *et al.*, 1975).

Thus the large size of polytene chromosomes allows a direct visual-ization of the transcriptional process and indicates that, as in other situations, gene activity in the salivary gland is regulated at the level of transcription.

Figure 3.10 The newly synthesized RNA made following ecdysone stimulation can be labeled with [³H]uridine and shown to hybridize to the puffs that form following ecdysone treatment. Conversely, a puff that regresses after hormone treatment binds only the labeled RNA synthesized before addition of the hormone.

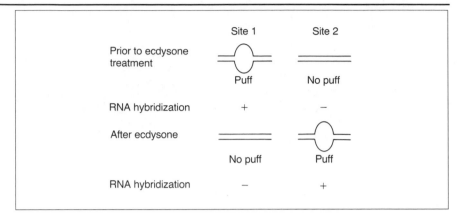

3.3 *Regulation at transcriptional elongation*

3.3.1 Initiation of transcription

In the vast majority of cases where increased transcription of a particular gene has been demonstrated, it is likely that such increased transcription is mediated by an increased rate of initiation of transcription by RNA polymerase which occurs at the region of the gene known as the promoter (Section 6.2.1). Hence in a tissue in which a gene is being transcribed actively, a large number of polymerase molecules will be moving along the gene at any particular time, resulting in the production of a large number of transcripts. Such a series of nascent transcripts being produced from a single transcription unit can be visualized in the lampbrush chromosomes of amphibian oocytes, the nascent transcripts associated with each RNA polymerase molecule increasing in length the further the polymerase has proceeded along the gene, resulting in the characteristic nested appearance (Figure 3.11).

By contrast, in tissues where a gene is transcribed at very low levels, initiation of transcription will be a rare event and only one or a very few polymerase molecules will be transcribing a gene at any particular time. Similarly, the absence of transcription of a particular gene in some tissues will result from a failure of RNA polymerase to initiate transcription in that tissue (Figure 3.12). A large number of sequences upstream of the point at which initiation occurs and which are involved in its regulation have now been described, and these will be discussed in Chapter 6.

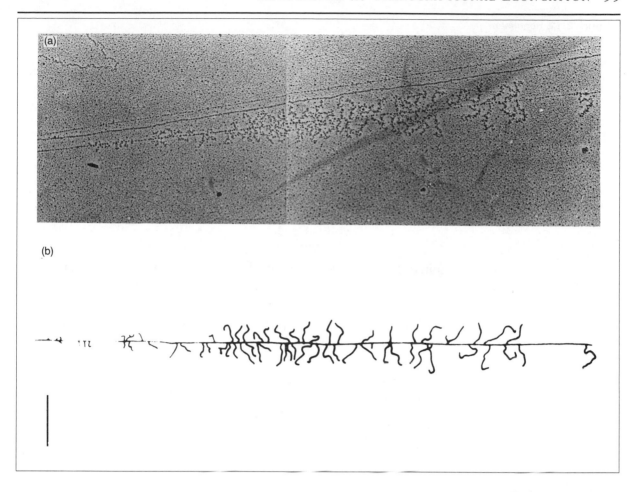

3.3.2 Polymerase pausing

Although the majority of cases of transcriptional regulation are likely to occur at the level of initiation, some cases have been described where regulation occurs following the initiation of transcription by RNA polymerase and the production of a short truncated RNA (for review, see Greenblatt *et al.*, 1993).

An example of this is seen in the case of the human immunodeficiency virus (HIV-1) promoter (for reviews, see Cullen, 1991; Greenblatt *et al.*, 1993). Thus early in infection with this virus, a low level of transcription occurs from this promoter and many of the transcripts terminate very close to the promoter producing very short RNAs. Subsequently the viral Tat protein binds to these RNAs at a region known as Tar which is located at +19 to +42 relative to the start site of transcription at nucleotide +1. This binding produces two effects. Firstly, there is a large increase in transcriptional initiation leading to many more RNA transcripts being

Figure 3.11 Electron micrograph (A) and summary diagram (B) of a lampbrush chromosome in amphibian oocytes, showing the characteristic nested appearance produced by the nascent mRNA chains attached to transcribing RNA polymerase molecules. The bar indicates 1 μm.

Figure 3.12 Regulation of transcriptional initiation results in differences in the number of RNA polymerase molecules transcribing a gene and therefore in the number of transcripts produced.

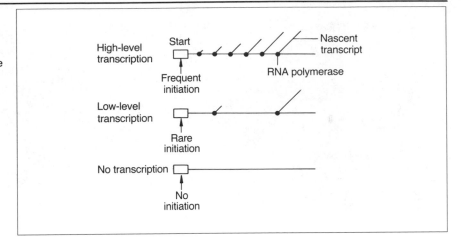

initiated. In addition, however, Tat also overcomes the block to elongation, leading to the production of a greater proportion of long transcripts which can encode the viral proteins.

Hence Tat produces a large increase in the production of HIV transcripts able to encode viral proteins both by stimulating transcriptional initiation and by overcoming polymerase pausing close to the promoter (Figure 3.13). It is likely that this regulation of polymerase pausing close to the gene promoter also operates for specific cellular genes. Indeed it has been suggested that it may be involved in the increased expression of the gene encoding the heat-inducible hsp70 protein that occurs following exposure of cells to elevated temperature and which results from the

Figure 3.13 The HIV Tat protein acts both by enhancing the rate of transcriptional initiation by RNA polymerase (RNA P) and by enhancing the rate of elongation by overcoming polymerase pausing close to the promoter. Both these effects are achieved by the Tat protein binding to a specific region (TAR) of the nascent RNA.

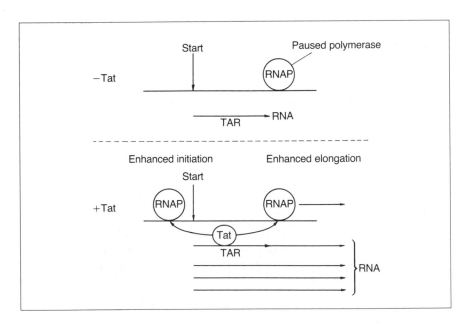

binding of the heat-shock transcription factor (HSF) to the promoter of this gene (Sections 3.2.4 and 6.2.1; for review, see Lis and Wu, 1993). Thus far, however, Tat is unique in that it exerts its effects on initiation and elongation by binding to RNA rather than by binding to DNA like HSF and all other known transcription factors.

3.3.3 Transcriptional elongation

As well as the release of polymerase which has paused close to the promoter, it is also possible for transcriptional regulation to operate by releasing a block to elongation of the RNA transcript which occurs at a considerable distance from the start site of transcription (for review, see Spencer and Groudine, 1990).

This form of regulation is responsible, for example, for the 10-fold decline in mRNA levels encoding the cellular oncogene c-*myc* (for discussion of cellular oncogenes see Chapter 8) which occurs when the human pro-myeloid cell line HL-60 is induced to differentiate into a granulocyte type cell. If nuclear run-on assays are carried out using nuclei from undifferentiated or differentiated HL-60 cells, the results obtained vary depending on the region of the c-*myc* gene whose transcription is being measured (Bentley and Groudine, 1986). Thus the c-*myc* gene consists of three exons, which appear in the mRNA and which are separated by intervening sequences that are removed from the primary transcript by RNA splicing. If the labeled products of the nuclear run-on procedure are hybridized to the DNA of the second exon, the levels of transcription observed are approximately 10-fold higher in nuclei derived from undifferentiated cells than in nuclei from differentiated cells. Hence the observed differences in c-*myc* RNA levels in these cell types are indeed produced by differences in transcription rates. However, if the same labeled products are hybridized to DNA from the first exon of the c-*myc* gene, virtually no difference in the level of transcription of this region in the differentiated compared with the undifferentiated cells is observed. Comparison of the rates of transcription of the first and second exons in undifferentiated and differentiated cells indicates that regulation takes place at the level of transcriptional elongation, rather than initiation. Thus, although similar numbers of polymerase molecules initiate transcription of the c-*myc* gene in both cell types, the majority terminate in differentiated cells near the end of exon 1, do not transcribe the remainder of the gene and hence do not produce a functional RNA. In contrast, in undifferentiated cells most polymerase molecules that initiate transcription, transcribe the whole gene and produce a functional RNA. Hence the fall in c-*myc* RNA in differentiated cells is regulated by means of a block to elongation of nascent transcripts

Figure 3.14 Regulation of transcriptional elongation in the c-*myc* gene. A block to elongation at the end of exon 1 results in most transcripts terminating at this point in differentiated HL-60 cells.

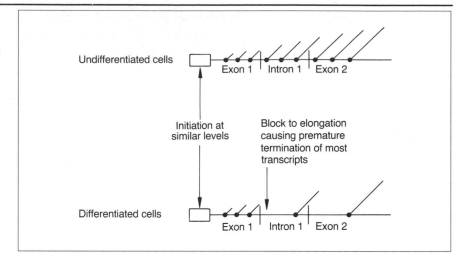

(Figure 3.14). A 180 bp sequence from the 3' end of the first exon of c-*myc* has been shown to mediate this effect and to block transcriptional elongation if placed within the transcribed region of another gene (Wright and Bishop, 1989). Interestingly, the rapid inhibition of transcriptional elongation following differentiation is supplemented, several days after differentiation, by an inhibition of c-*myc* transcription at the level of initiation.

Similar effects on transcriptional elongation have been seen in several other cellular oncogenes such as c-*myb*, c-*fos* and c-*mos*, indicating that this mechanism is not confined to a single gene and may be quite widespread (for review, see Spencer and Groudine, 1990).

Thus in different situations, enhanced transcription can arise from increased initiation, release of paused polymerase or increased elongation (Figure 3.15). In general, however, it is likely that in most cases transcription is regulated at the level of initiation, with specific regulatory processes acting to enhance the rate at which RNA polymerases initiate transcription of the DNA into RNA. As with DNA amplification or rearrangement (Chapter 2) control at polymerase release or elongation is likely to represent a response to the requirements of a particular situation. Thus these processes may, for example, allow a rapid response to a particular situation since the RNA polymerase and associated proteins are already poised on the gene rather than having to bind and initiate transcription *de novo*.

As well as having in common the ability to respond rapidly to the need for new transcription, both the release of paused polymerase and increased elongation resemble one another in that they both involve the further movement of polymerase which has stalled on the gene although they differ in whether such stalling occurs adjacent to or at a distance from the gene promoter. It is possible therefore that these two processes may have a

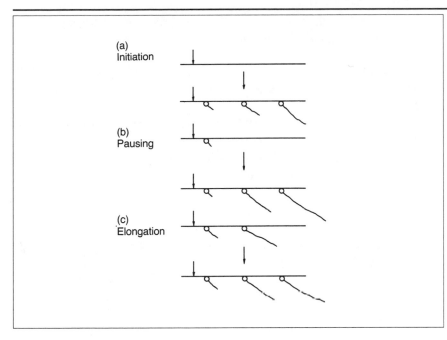

Figure 3.15 Summary of the mechanisms by which transcription can be enhanced, i.e. increased initiation (a), release of paused polymerase (b) or enhanced elongation (c).

similar fundamental mechanism, perhaps involving direct modification of the RNA polymerase itself to allow it to pass the block (for discussion, see Spencer and Groudine, 1990). At present, however, the mechanisms mediating these effects remain obscure and it is unclear whether the same processes are involved in each case.

3.4 Conclusions

The experiments described in this chapter provide conclusive evidence that control of transcriptional initiation is the primary means used to regulate gene expression in eukaryotic organisms. The mechanism by which such transcriptional control is achieved will be discussed in Chapters 5, 6 and 7. Some cases of post-transcriptional control have been described, however, and these will be discussed in the next chapter.

References

Bentley, D. L. and M. Groudine 1986. A block to elongation is largely responsible for decreased transcription of c-*myc* in differentiated HL 60 cells. *Nature* **321**, 702–6.

Bonner, J. J. and M. L. Pardue 1977. Ecdysone-stimulated RNA synthesis in salivary glands of *Drosophila melanogaster* assay by *in situ* hybridization. *Cell* **12**, 219–25.

Cullen, B. R. 1991. Regulation of HIV-1 gene expression. *FASEB Journal* **5**, 2361–8.

Darnell, J. E. 1982. Variety in the level of gene control in eukaryotic cells. *Nature* **297**, 365–71.

Davidson, E. H. and R. J. Britten 1979. Regulation of gene expression: possible role of repetitive sequences. *Science* **204**, 1052–9.

Derman, E., K. Krauter, L. Walling, C. Weinberger, M. Ray and J. E. Darnell, Jr 1981. Transcriptional control in the production of liver specific mRNAs. *Cell* **23**, 731–9.

Gilmour, R. S., P. R. Harrison, J. D. Windass, N. A. Affara and J. Paul 1974. Globin messenger RNA synthesis and processing during haemoglobin induction in Friend cells. 1. Evidence for transcriptional control in clone M2. *Cell Differentiation* **3**, 9–22.

Goldberg, R. B., G. Hosheck, G. S. Ditta and R. W. Breidenbach 1981. Developmental regulation of cloned superabundant mRNAs in soybean. *Developmental Biology* **83**, 218–31.

Greenblatt, J., J. R. Nodwell and S. W. Mason 1993. Transcriptional antitermination. *Nature* **364**, 401–6.

Groudine, M., H. Hoitzer, K. Scherner and A. Therwath 1974. Lineage dependent transcription of globin genes. *Cell* **3**, 243–7.

Groudine, M., M. Peretz and H. Weintraub 1981. Transcriptional regulation of hemoglobin switching in chicken embryos. *Molecular and Cellular Biology* **1**, 281–8.

Jacob, F. and J. Monod 1961. Genetic regulatory mechanisms in the synthesis of proteins. *Journal of Molecular Biology* **3**, 318–56.

Jeffreys, A. J. and R. A. Flavell 1977. The rabbit β-globin gene contains a large insert in the coding sequence. *Cell* **12**, 1097–108.

Kamaly, J. C. and R. B. Goldberg 1980. Regulation of structural gene expression in tobacco. *Cell* **19**, 935–46.

Lamond, A. I. 1991. Nuclear RNA processing. *Current Opinion in Cell Biology* **3**, 493–501.

Latchman, D. S., H. Brzeski, R. H. Lovell-Badge and M. J. Evans 1984. Expression of the alpha-foetoprotein gene in pluripotent and committed cells. *Biochimica et Biophysica Acta* **783**, 130–6.

Lis, J. and C. Wu 1993. Protein traffic on the heat shock promoter: parking stalling and trucking along. *Cell* **74**, 1–4.

Lowenhaupt, K., C. Trent and J. B. Lingrel 1978. Mechanisms for accumulation of globin mRNA during dimethyl sulfoxide induction of mouse erythroleukaemia cells: synthesis of precursors and mature mRNA. *Developmental Biology* **63**, 441–54.

Nevins, J. R. 1983. The pathway of eukaryotic mRNA transcription. *Annual Review of Biochemistry* **52**, 441–6.

Powell, D. J., J. M. Freidman, A. J. Oulethe, K. S. Krauter and J. E. Darnell, Jr 1984. Transcriptional and post-transcriptional control of specific messenger RNAs in adult and embryonic liver. *Journal of Molecular Biology* 179, 21–35.

Roop, D. R., J. L. Nordstrom, S.-Y. Tsai, M.-J. Tsai and B. W. O'Malley 1978. Transcription of structural and intervening sequences in the ovalbumin gene and identification of potential ovalbumin mRNA precursors. *Cell* 15, 671–85.

Sharp, P. A. 1987. Splicing of messenger RNA precursors. *Science* 235, 766–71.

Spencer, C.A. and Groudine, M. 1990. Transcriptional elongation and eukaryotic gene regulation. *Oncogene* 5, 777–85.

Spradling, A., S. Penman and M. L. Pardue 1975. Analysis of *Drosophila* mRNA by *in situ* hybridization: sequences transcribed in normal and heat shocked cultured cells. *Cell* 4, 395–404.

Swaneck, G. E., J. L. Nordstrom, F. Kreuzaler, M.-J. Tsai and B. W. O'Malley 1979. Effect of estrogen on gene expression in chicken oviduct, evidence for transcriptional control of the ovalbumin gene. *Proceedings of the National Academy of Sciences of the USA* 76, 1049–53.

Tilghman, S. M., P. J. Curtis, D. C. Tiemeier, P. Leder and C. Weissmann 1978. The intervening sequence of a mouse beta-globin gene is transcribed within the 15S beta-globin mRNA precursor. *Proceedings of the National Academy of Sciences of the USA* 75, 1309–13.

Wold, B. J., W. H. Klein, B. R. Hough-Evans, R. J. Britten and E. H. Davidson 1978. Sea urchin embryo mRNA sequences expressed in the nuclear RNA of adult tissues. *Cell* 14, 941–50.

Wright, S. and J. M. Bishop 1989. DNA sequences that mediate attenuation of transcription from the mouse proto-oncogene c-*myc*. *Proceedings of the National Academy of Sciences of the USA* 86, 505–9.

4 Post-transcriptional regulation

4.1 Regulation after transcription?

Although the evidence discussed in the preceding chapter indicates that, in mammals at least, the primary control of gene expression lies at the level of transcription, a number of cases exist where changes in the rate of synthesis of a particular protein occur without a change in the transcription rate of the corresponding gene. Indeed, in some lower organisms post-transcriptional regulation may constitute the predominant form of regulation of gene expression.

In the sea urchin, for example, the nuclear RNA contains many more different RNA species than are found in the cytoplasmic mRNA. Hence a large proportion of the genes transcribed give rise to RNA products that are not transported to the cytoplasm and do not function as a mRNA. Interestingly, however, this process is regulated differently in different tissues, an RNA species which is confined to the nucleus in one tissue being transported to the cytoplasm and functioning as a mRNA in another tissue. Thus, up to 80% of the cytoplasmic mRNAs found in the embryonic blastula are absent from the cytoplasmic RNA of adult tissues, such as the intestine, but are found in the nuclear RNA of such tissues (Wold *et al.*, 1978).

Although post-transcriptional regulation in mammals does not appear to be as generalized as in the sea urchin, some cases exist where changes in cytoplasmic mRNA levels occur without alterations in the rate of gene transcription. Such post-transcriptional regulation may be more important in controlling variations in the level of mRNA species expressed in all tissues than in the regulation of mRNA species that are expressed in only one or a few tissues. For example, in the experiments of Darnell and colleagues (Powell *et al.*, 1984), which demonstrated the importance of transcriptional control in the regulation of liver-specific mRNAs (Section 3.2.3), tissue-specific differences in the levels of the mRNAs encoding actin and tubulin (which are expressed in all cell types) were observed in the absence of differences in transcription rates. Clearly, these and other cases where mRNA levels alter in the absence of changes in transcription rates indicate the existence of post-transcriptional control processes and require an understanding of their mechanisms.

In principle, such post-transcriptional regulation could operate at any of the many stages between gene transcription and the translation of the corresponding mRNA in the cytoplasm. Indeed, the available evidence indicates that in different cases regulation can occur at any one of these levels. Each of these will now be discussed in turn.

4.2 Regulation of RNA splicing

4.2.1 RNA splicing

The finding that the protein-coding regions of eukaryotic genes are split by intervening sequences (introns) which must be removed from the initial transcript by RNA splicing of the protein-coding exons (Figure 4.1; for review, see Sharp 1987; Lamond, 1991a) led to much speculation that this process might provide a major site of gene regulation. Thus, in theory, an RNA species transcribed in several tissues might be correctly spliced to yield functional RNA in one tissue and remain unspliced in another tissue. An unspliced RNA would either be degraded within the nucleus or, if transported to the cytoplasm, would be unable to produce a functional protein due to the interruption of the protein-coding regions (Figure 4.2). Several cases of such processing versus discard decisions have now been described in *Drosophila* (reviewed by Bingham *et al.*, 1988). One decision of this type is involved in regulating the movement of the transposable P element in *Drosophila*. This DNA element encodes a protein which allows it to move from position to position within the genome, but this process occurs only in the germ cells and not in somatic

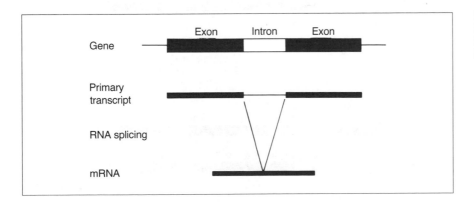

Figure 4.1 Removal of an intervening sequence from the primary RNA transcript by RNA splicing.

Figure 4.2 The absence of RNA splicing of a transcript in a particular tissue results in a lack of production of the corresponding protein.

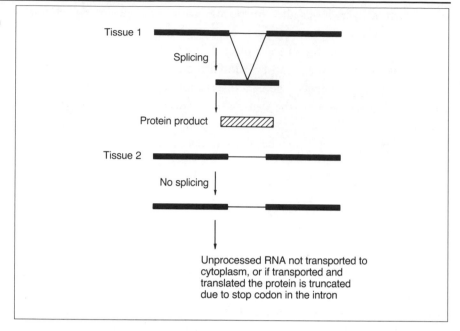

tissue. Detailed analysis of the gene encoding this protein (Figure 4.3) has shown that to produce a functional mRNA three intervening sequences must be removed from the initial transcript. The third of these intervening sequences is removed only in germ cells so that somatic cells accumulate a non-functional, partially spliced transcript in which only the first two intervening sequences have been removed (for review, see Rio, 1991).

4.2.2 Alternative RNA splicing

Although no clear case of a processing versus discard decision has as yet been defined in mammals, numerous cases of alternative RNA splicing

Figure 4.3 Removal of intron 3 of the P element transposase transcript occurs only in germ line tissue.

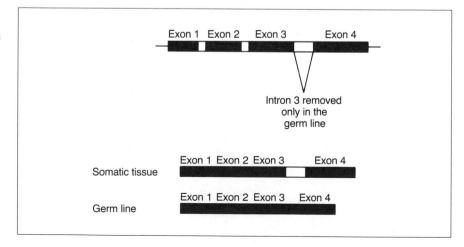

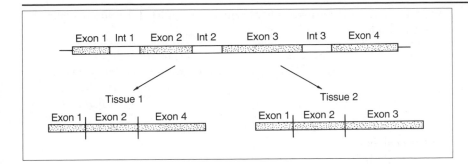

Figure 4.4 Alternative splicing of the same primary transcript in two different ways results in two different mRNA molecules.

have been described, both in mammals and other organisms. In this process (for reviews, see Latchman, 1990; McKeown, 1992) a single gene is transcribed in several different tissues, the transcripts from this gene being processed differentially to yield different functional mRNAs in the different tissues (Figure 4.4). In many cases these RNAs are translated to yield different protein products. It is noteworthy that this mechanism of gene regulation involves not only regulation of processing but also regulation of transcription, in that the alternatively processed RNAs are transcribed in only a restricted range of cell types and not in many other cells.

Cases of alternative RNA processing occur in the genes involved in a wide variety of different cellular processes, ranging from genes which regulate embryonic development or sex determination in *Drosophila* to those involved in muscular contraction or neuronal function in mammals. A representative selection of such cases is given in Table 4.1.

For convenience, cases of alternative RNA processing can be divided into three groups (Leff *et al.*, 1986):

1. Situations where the 5' end of the differentially processed transcripts is different.
2. Situations where the 3' end of the differentially processed transcripts is different.
3. Situations where both the 5' and 3' ends of the differentially processed transcripts are identical.

Situations where the 5' end of the transcripts is different

In these cases, two alternative primary transcripts are produced by transcription from different promoter elements and these are then processed differentially. In several situations, such as the mouse α-amylase gene (Figure 4.5), differential splicing is controlled simply by the presence or absence of a particular exon in the primary transcript. Thus in the salivary gland, where transcription takes place from an upstream promoter, the exon adjacent to this promoter is included in the processed RNA and a

Table 4.1 Cases of alternative splicing which are regulated developmentally or tissue specifically

Protein	Species	Nature of transcripts which undergo alternative splicing	Cell types carrying out alternative splicing
(a) *Immune system* Immunoglobulin heavy-chain IgD, IgE, IgG, IgM	Mouse	3' end differs	B cells
Lyt-2	Mouse	Same transcript	T cells
(b) *Enzymes* Alcohol dehydrogenase	*Drosophila*	5' end differs	Larva and adult
Aldolase A	Rat	5' end differs	Muscle and liver
α-Amylase	Mouse	5' end differs	Liver and salivary gland
(2'5') oligo A synthetase	Human	3' end differs	B cells and monocytes
(c) *Muscle* Myosin light chain	Rat/mouse/ human/chicken	5' end differs	Cardiac and smooth muscle
Myosin heavy chain	*Drosophila*	3' end differs	Larval and adult muscle
Tropomyosin	Mouse/rat/ human/ *Drosophila*	Same transcript	Different muscle cell types
Troponin T	Rat/quail/ chicken	Same transcript	Different muscle cell types
(d) *Nerve cells* Calcitonin/CGRP	Rat/human	3' end differs	Thyroid C cells or neural tissue
Myelin basic protein	Mouse	Same transcript	Different glial cells
Neural cell adhesion molecule	Chicken	Same transcript	Neural develop- ment
Preprotachykinin	Bovine	Same transcript	Different neurons
(e) *Others* Fibronectin	Rat/human	Same transcript	Fibroblasts and hepatocytes
Early retinoic acid- induced gene 1	Mouse	Same transcript	Stages of embryonic cell differentiation
Thyroid hormone receptor	Rat	Same transcript	Different tissues

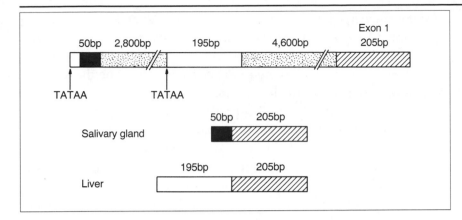

Figure 4.5 Alternative splicing at the 5' end of α-amylase transcripts in the liver and salivary gland. The two alternative start sites for transcription are indicated (TATAA) together with the 5' region of the mRNAs produced in each tissue.

downstream exon is omitted. In the liver, where the transcripts are initiated 2.8 kb downstream and do not contain the upstream exon, the processed RNA includes the downstream exon (Young *et al.*, 1981). However, other cases of this type, for example that of the myosin light-chain gene (Figure 4.6), are more complex with each of the alternative primary transcripts containing both the alternatively spliced exons. In such cases it is assumed that the different primary transcripts fold into different secondary structures which favor the different splicing events.

Whether this is the case or not, it is clear that cases of alternative splicing arising from differences in the site of transcriptional initiation represent further examples of transcriptional regulation in which the variation in RNA splicing is secondary to the selection of the different promoters in the different tissues. This is not so for the remaining two categories of alternative processing event.

Situations where the 3' end of the transcripts is different
After the primary transcript has been produced, it is rapidly cleaved at a

Figure 4.6 Alternative splicing of the myosin light chain transcripts in different muscle cell types produces two mRNAs (1F and 3F) differing at their 5' ends. The two alternative start sites of transcription used to produce each of the RNAs are indicated (TATAA), together with the intron–exon structure of the gene.

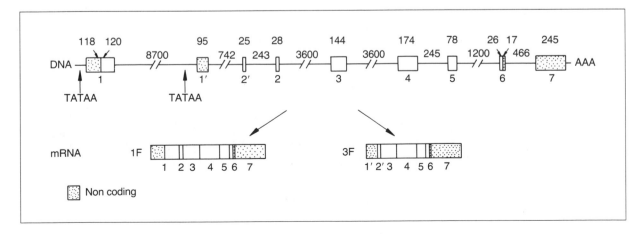

point downstream of the protein-coding information and a run of adenosine residues (the poly A tail) is added post-transcriptionally. In many genes the process of cleavage and polyadenylation occurs at a different position within the primary transcript in different tissues and the different transcripts are then differentially spliced.

The best-defined example of this process occurs in the genes encoding the immunoglobulin heavy chain of the antibody molecule and plays an important role in the regulation of the antibody response to infection. Thus, early in the immune response, the antibody-producing B cell synthesizes membrane-bound immunoglobulin molecules whose interaction with antigen triggers proliferation of the B cell and results in the production of more antibody-synthesizing cells. The immunoglobulin produced by these cells is secreted, however, and can interact with antigen in tissue fluids, triggering the activation of other cells in the immune system. The production of membrane-bound and secreted immunoglobin molecules is controlled by the alternative splicing of different RNA molecules differing in their 3' ends (Figure 4.7). The longer of these two molecules contains two exons encoding the portion of the protein that anchors it in the membrane. When this molecule is spliced, both these two exons are included, but a region encoding the last 20 amino acids of the secreted form is omitted. In the shorter RNA, the two transmembrane domain-

Figure 4.7 Alternative splicing of the immunoglobulin heavy chain transcript at different stages of B cell development. The two unspliced RNAs produced by use of the two alternative polyadenylation sites in the gene are shown, together with the spliced mRNAs produced from them.

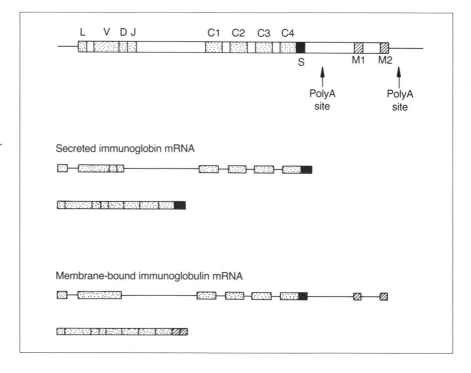

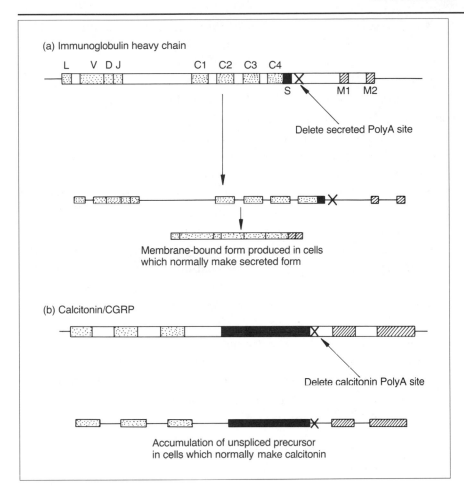

(a) Immunoglobulin heavy chain

Delete secreted PolyA site

Membrane-bound form produced in cells
which normally make secreted form

(b) Calcitonin/CGRP

Delete calcitonin PolyA site

Accumulation of unspliced precursor
in cells which normally make calcitonin

Figure 4.8 Effect of deleting the more upstream of the two polyadenylation sites in the immunoglobulin heavy chain (A) and calcitonin/CGRP genes (B) on the production of the alternatively spliced RNAs derived from each of these genes.

encoding exons are absent and the region specific to the secreted form is included in the final mRNA.

If the polyadenylation site used in the production of the shorter immunoglobulin RNA is artificially removed, preventing its use (Danner and Leder, 1985), the expected decrease in the production of secreted immunoglobulin is paralleled by a corresponding increase in the synthesis of the membrane-bound form of the protein (Figure 4.8A). This indicates that the choice of splicing pattern is controlled by which polyadenylation site is used, removal of the upstream site resulting in increased use of the downstream site and increased production of the mRNA encoding the membrane-bound form.

This finding indicates that in at least some cases of this type the primary regulatory event to be understood is that determining the site of cleavage and polyadenylation, and that, as with cases of differential promoter usage, alternative RNA splicing is regulated by differences in the structure of the transcript produced in different tissues.

Not all cases of alternative RNA splicing where the 3' end of the alternatively spliced RNAs varies are of this type, however. This conclusion has emerged from the intensive study of the gene encoding the calcium regulatory protein, calcitonin, carried out by Rosenfeld and colleagues (Rosenfeld *et al.*, 1984; Leff *et al.*, 1988). When the gene encoding the calcitonin protein (which is a small peptide of 32 amino acids) was isolated it was found that it had the potential to produce an RNA encoding an entirely different peptide of 36 amino acids, which was named calcitonin-gene-related peptide (CGRP). Unlike calcitonin, which is produced in the thyroid gland, CGRP is produced in specific neurons within the brain and peripheral nervous system. These two peptides are produced (Figure 4.9) by alternative splicing of two distinct transcripts differing in their 3' ends.

This case differs from that of the immunoglobulin heavy chain, however, in that deletion of the polyadenylation site used in the shorter, calcitonin-encoding RNA does not result in an increase in CGRP expression in cells normally expressing calcitonin. Instead, large unspliced transcripts utilizing the downstream (CGRP) polyadenylation site accumulate in these cells (Figure 4.8B). Although in CGRP-producing cells such transcripts would normally be spliced to yield CGRP mRNA, this does not occur in these experiments in cells normally producing calcitonin, and hence these unspliced precursors accumulate. This shows that in the calcitonin/CGRP gene the use of different polyadenylation sites is

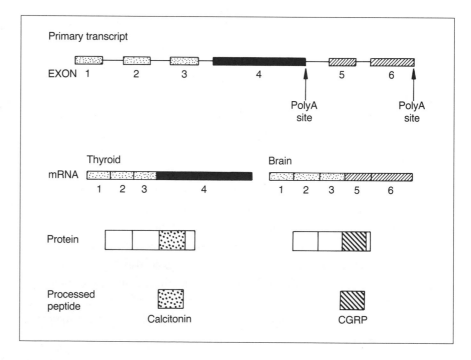

Figure 4.9 Alternative splicing of the calcitonin/CGRP gene in brain and thyroid cells. Alternative splicing followed by proteolytic cleavage of the protein produced in each tissue yields calcitonin in the thyroid and CGRP in the brain.

secondary to the difference in RNA splicing, suggesting the existence of tissue-specific splicing factors, whose presence or absence in a specific tissue determines the pattern of calcitonin/CGRP RNA splicing. In particular, a specific factor present in CGRP-producing tissues would be required for CGRP-specific splicing, while in its absence such splicing could not occur. Hence, when the calcitonin-specific polyadenylation site is removed, no CGRP-specific splicing can occur in tissues normally producing calcitonin and unspliced precursors accumulate. The nature of such factors and the mechanism by which they act are discussed below.

Situations where both the 5' and 3' ends of the differently processed transcripts are identical
The existence of tissue-specific splicing factors which regulate alternative splicing is also indicated by the existence of cases where a transcript with identical 5' and 3' ends is spliced differently in different tissues, and which therefore cannot be explained by differential usage of promoters or polyadenylation sites.

Although the initial reports of such cases were confined to the eukaryotic DNA viruses, such as SV40 and adenovirus, a number of cases involving the cellular genes of higher eukaryotes have now been described. One of the most dramatic of these described so far (Figure 4.10) involves the skeletal muscle troponin T gene (Breitbart *et al.*, 1987) in which the same RNA can be spliced in up to 64 different ways in different muscle cell types. The existence of tissue-specific splicing factors acting on this gene is indicated by the finding that the artificial introduction and expression of this gene in non-muscle cells or myoblasts results in the complete removal of exons 4–8, whereas in muscle cells (myotubes) the correct pattern of alternative splicing seen with the endogenous gene is reproduced faithfully.

4.2.3 Mechanism of alternative RNA splicing

Evidence for the existence of alternative splicing factors
The mechanism of alternative RNA splicing has been analyzed in most detail in the case of the calcitonin/CGRP gene referred to above. As already discussed, the observation that deletion of the calcitonin-specific polyadenylation site does not result in increased production of CGRP mRNA in tissues normally producing calcitonin suggests that particular splicing factor(s) expressed only in CGRP-producing tissues are required for the production of the CGRP mRNA.

The possible existence of such factors has been further investigated by removing the promoter of the calcitonin/CGRP gene (which functions

Figure 4.10 Alternative splicing of the four combinatorial exons (4–8) and the two mutually exclusive exons (16 and 17) can result in up to 64 distinct mRNAs from the rat troponin T gene.

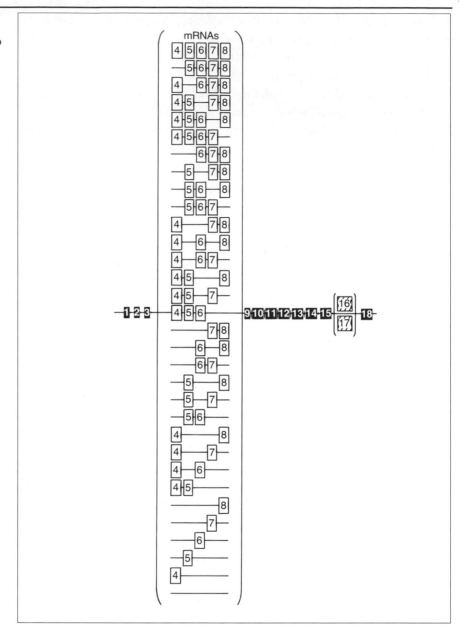

only in a restricted range of cells) and replacing it with the metallothionein promoter, which functions in all cell types. This construct was then introduced into a fertilized mouse egg *in vitro*. Two-cell embryos developing from these eggs were re-implanted into foster mothers and allowed to develop. This resulted in the eventual production of a transgenic animal in which every cell was expressing the calcitonin/CGRP gene. In such animals the vast majority of tissues produced only the spliced calcitonin

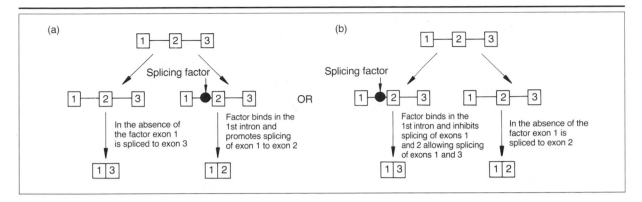

Figure 4.11 Possible models by which an alternative splicing factor can affect splicing by binding to a *cis*-acting sequence. In (A), the factor acts by binding to the weaker of the two potential splicing sites, promoting its use, while in (B) it acts by binding to the stronger of the two sites and inhibiting its use so that the other, weaker, site is used.

mRNA, while only heart and brain (the natural site of CGRP production) produced the spliced CGRP mRNA.

These experiments suggested that although all tissues have the capacity to produce calcitonin mRNA, an extra factor found only in some tissues is required to produce the CGRP mRNA. This factor would obviously be present in the CGRP-producing brain tissue. Its apparent presence in the heart, which does not normally transcribe the calcitonin/CGRP gene, would be explained on the basis of its normal participation in other alternative splicing events involving genes expressed in the heart. Thus a small number of alternative splicing factors could control many different alternative splicing events in the tissues which express them. In the presence of such a factor one particular pattern of splicing would occur, whereas in its absence the other, default pathway would be followed.

Mode of action of alternative splicing factors

That certain factors are necessary for each particular pattern of splicing clearly begs the question of how such factors act. It is likely that these factors recognize *cis*-acting sequences within the RNA transcript itself. Clearly, the interaction of these factors with such sequences could produce alternative splicing either by promoting splicing at the site of the *cis*-acting sequence at the expense of the alternative splice site (Figure 4.11A) or by inhibiting splicing at the site of binding and thereby promoting the use of the alternative splice site (Figure 4.11B).

Both of these types of mechanism appear to be used in different cases. Indeed examples of each of these mechanisms can be seen in the hierarchy of alternatively spliced genes which regulates sex determination in *Drosophila* (for review, see Baker, 1989). In this hierarchy each gene product controls the alternative splicing of the next gene in the pathway resulting in different protein products in males and females. In turn these products differentially regulate the splicing of the next gene in the

Figure 4.12 Schematic diagram (not to scale) of the hierarchy of alternatively spliced genes which controls sex determination in *Drosophila*. Female-specific splicing of the *Sxl* transcript produces a protein which promotes the female-specific splicing of its own RNA and that of the *tra* gene. In turn, the protein produced by the *tra* female-specific transcript in conjunction with the *tra*-2 gene product promotes the female-specific splicing of the *dsx* transcript. The sites of mutations in the *Sxl* and *dsx* genes which affect their sex-specific splicing are indicated. Modified from Baker (1989).

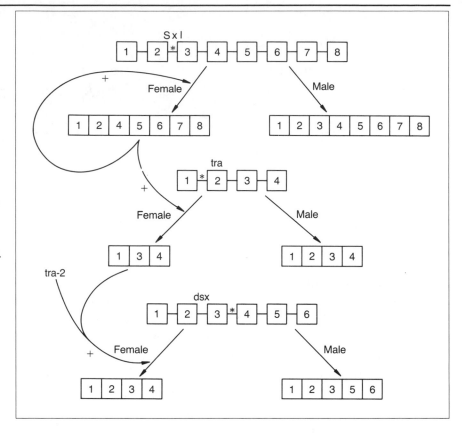

hierarchy, leading ultimately to the production of a male or female fly (Figure 4.12).

Thus the *Sxl* gene is differentially spliced in males and females, with the product of the female-specific mRNA not only controlling the splicing of the next gene in the hierarchy, *tra*, but also promoting its own female-specific splicing. Mutations which affect this autocatalytic function of *Sxl* on its own RNA map in the intron between exon 2 and the male-specific exon 3, indicating that the SxL protein acts by inhibiting the male-specific splicing of exons 2 and 3. Similarly *Sxl* prevents the use of the male-specific exon 2 in the *tra* gene by binding within intron 1 and preventing the binding of the constitutive splicing factor U2AF which is essential for removal of all introns (Valcarcel *et al.*, 1993).

In contrast, the action of the tra and tra-2 gene products on the splicing of the *dsx* transcript appears to be mediated by promoting the use of the female-specific splice rather than the constitutive male-specific splicing event. In this case, mutations in *dsx* which affect its splicing map in the intron between exon 3 and the female-specific exon 4, indicating that the alternative splicing factor binds here and promotes the splicing of exons 3 and 4.

Nature of alternative splicing factors
The evidence for the existence of alternative splicing factors and the identification of *cis*-acting sequences with which they interact have led to many
attempts to identify these factors. These have involved either genetic
approaches or approaches based on studying the expression patterns of
splicing factors.

Genetic studies
In *Drosophila*, which is genetically very well characterized, the main
approach to this problem has been a genetic one. Thus as discussed above,
genes such as *Sxl* and *tra*, originally identified by the fact that mutations
within them disrupted the process of sex determination, are now known to
act by controlling alternative splicing. The products of these genes are
alternative splicing factors and their study allows a unique insight into the
nature of such factors.

The sequencing of the *Sxl* and *tra-2* genes has revealed that the proteins
they encode contain one or more copies of a ribonucleoprotein (RNP)
consensus sequence that is found in a wide variety of RNA-binding
proteins, such as those of the mammalian spliceosome, and is likely to
constitute an RNA-binding domain (reviewed by Bandziulis *et al.*, 1989).
Hence SxL and Tra-2 are likely to influence alternative splicing by binding
directly to the *cis*-acting sequences in the spliced RNAs discussed above.

Tissue-specific splicing factors
In mammals and other vertebrates where the genetic approach is not
as useful as in *Drosophila*, the search for alternative splicing factors has
mainly concentrated on the identification of tissue-specific variants of the
constituents of the multicomponent spliceosome which catalyzes RNA
splicing. This structure contains both proteins and small RNA molecules
ranging from 56 to 217 bases in size and known as the U RNAs (reviewed
by Maniatis and Reed, 1987). Although the majority of proteins and
RNAs of the spliceosome are apparently expressed in all tissues, variants of
the U1 RNA which are expressed only in certain tissues or stages of development have been reported in both mammals and Amphibia (see, e.g.
Lund *et al.*, 1985). However, the expression of a specific variant of U1 in
certain tissues has not thus far been correlated with a particular pattern of
alternative RNA splicing in these tissues.

However, such a correlation has been reported in the case of a protein
component of the spliceosome known as SmN (for review, see Latchman,
1990). This protein is closely related to the constitutively expressed splicing protein SmB but its expression amongst mouse tissues is confined to

brain and heart; precisely the tissues which, in the experiments described above, were able to produce correctly spliced CGRP mRNA. This correlation suggested that SmN might be the tissue-specific factor which regulates the alternative splicing of the calcitonin/CGRP gene. However, the artificial expression of SmN in fibroblast cells, which can normally only carry out calcitonin-specific splicing, did not confer on them the ability to carry out CGRP-specific splicing (Delsert and Rosenfeld, 1992). Hence if SmN is indeed required for CGRP-specific splicing, other factors which are present in brain and heart tissue but absent in fibroblast cells must also be required for this to occur. Interestingly the absence of a functional SmN protein results in the death of mice shortly after birth (Section 5.7.2) indicating that SmN does fulfil an essential function presumably in the regulation of specific alternative splicing events in the brain and/or heart (Davies, 1992).

QUANTITATIVE VARIATIONS IN CONSTITUTIVELY EXPRESSED SPLICING FACTORS

The lack of direct evidence relating a particular vertebrate tissue-specific splicing factor to the regulation of a specific alternative splicing event has led to the suggestion that many tissue-specific splicing events might depend on quantitative variations in the levels of splicing factors which are present in all tissues. Thus the SF2 factor is a constitutively expressed protein which is present in all cells and is essential for the basic process of splicing itself. However, its concentration has been shown to influence which of two competing upstream splice sites is joined to a downstream site. Thus high concentrations of SF2 favor the more proximal of the two sites whilst low concentrations favor the more distal site (Figure 4.13; Krainer *et al.*, 1990; for review, see Lamond, 1991b). Interestingly, another constitutively expressed factor, hnRNPA1, has the opposite effect, favoring the use of the more distal site (Mayeda and Krainer, 1992). Hence, in this situation, the outcome of a specific alternative splicing event could be different in two different tissues depending on the relative concentration of SF2 and hnRNPA1 in each tissue even though both factors were present in both tissues (Figure 4.13).

Such a system in which the different outcome of alternative splicing events in each tissue is controlled by quantitative differences in the relative levels of constitutively expressed factors can readily be fitted into the mechanistic models illustrated in Figure 4.11 simply by suggesting that the two factors have opposite effects on the relative strengths of the two competing splice sites. Interestingly, like the SxL and TrA proteins discussed above, both SF2 and hnRNPA1 have the RNP consensus sequence

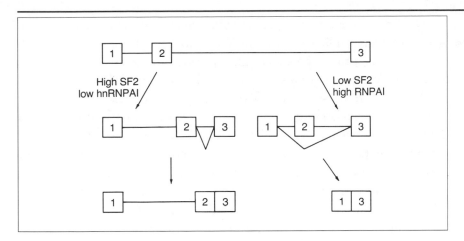

Figure 4.13 The pattern of splicing can be regulated by the balance in levels between the constitutively expressed SF2 and hnRNPA1 proteins. A high ratio of SF2 to hnRNPA1 favors use of the proximal exon (2) whilst a low ratio favors the distal exon (1).

found in RNA-binding proteins (Bandziulis *et al.*, 1989), indicating that they could act by binding directly to the competing splice sites and influencing their relative strength.

In some cases therefore alternative splicing need not involve factors expressed only in a limited range of tissues but may be dependent on quantitative variations in the levels of essential splicing factors which are expressed in all tissues. Indeed even in the case of the calcitonin/CGRP gene which was thought to clearly demonstrate the existence of a tissue-specific splicing factor, it has recently been suggested that quantitative rather than qualitative variations in the expression of different factors in the brain and heart compared with all other tissues might lead to the observed tissue-specific splicing pattern (Yeakley *et al.*, 1993). Hence the inability of tissues other than the brain and heart to carry out CGRP-specific splicing might be due to the absence of a tissue-specific splicing factor or to quantitative differences in the levels of constitutive splicing factors present in all tissues. In general, therefore, some alternative splicing events are likely to be regulated by qualitative changes in which specific splicing factors are present only in one sex or in one tissue whilst others may be regulated by quantitative changes in which constitutively expressed factors are present at different levels in different tissues.

4.2.4 Generality of alternative RNA splicing

The cases discussed above indicate the use of alternative splicing in a wide variety of biological processes. In mammals such splicing has been shown to regulate the immune system's production of antibodies, the production of neuropeptides such as CGRP and the tachykinins, substance P and substance K, as well as the synthesis of the different forms of at least four of the eight major sarcomere muscle proteins. Similarly, in *Drosophila* much of the posterior body plan is determined by developmentally regulated differential splicing of the ultrabithorax gene, while sex determination is also controlled by differential splicing of a hierarchy of genes in males and females (Section 4.2.3; reviewed by Baker, 1989).

The widespread use of alternative splicing in mammals does not refute the initial conclusion that gene regulation occurs primarily at the level of transcription, however. Thus, alternative splicing represents a response to a requirement for the production of related but different forms of a gene product in different tissues. It therefore supplements the regulation of transcription of the gene responsible for producing the different forms. Thus the immunoglobulin heavy-chain gene, which produces both membrane-bound and secreted forms of the protein at different stages of B cell development, is transcribed only in B cells and not in other cell types, while the transcription of the troponin T gene, which produces multiple different isoforms in different muscle cell types, is confined to differentiated muscle cells.

Interestingly, alternative splicing is particularly common in non-dividing cells such as nerve cells and muscle cells. As discussed in Chapter 5, the reprogramming of cellular commitment and gene transcription often requires a cell division event allowing alterations in the chromatin structure of the DNA and its associated proteins to occur. Hence alternative splicing may represent a means of conveniently changing the pattern of protein production in cells where such reprogramming cannot be achieved by cell division. Alternative splicing therefore represents a significant supplement to the regulation of transcription and the flexibility it confers suggests that many more cases will be identified in the future.

4.2.5 RNA editing

The finding that two different protein products can be produced from the same RNA by alternative splicing has been supplemented by the observation that a similar result can be achieved by a post-transcriptional sequence change in the mRNA (for review, see Hodges and Scott, 1992). Thus apolipoprotein B, which plays an important role in lipid transport, is known to exist in two closely related forms. A large protein of 512 kDa known as apo-B100 is synthesized by the liver, while a smaller protein apo-B48 is made by the intestine. The smaller protein is identical to the N-terminal portion of the larger protein. Analysis of the mRNA encoding these proteins revealed a 14.5 kb RNA in both tissues. These two RNAs were identical with the exception of a single base at position 6666, which is a cytosine in the liver transcript and a uridine in the intestinal transcript (Figure 4.14). This change has the effect of replacing a CAA codon, which

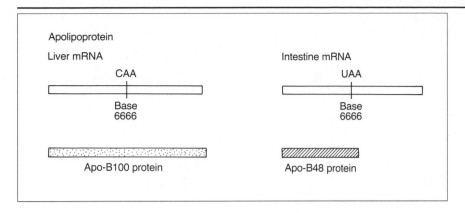

Figure 4.14 RNA editing of the apolipoprotein B transcript in the intestine produces an mRNA encoding the truncated protein apo-B48.

directs the insertion of a glutamine residue, with a UAA stop codon, which causes termination of translation of the intestine RNA and hence results in the smaller protein being made.

Only one gene encoding these proteins is present in the genome, and it is not alternatively spliced. In both intestinal and liver DNA this gene has a cytosine residue at position 6666. Hence the uridine in the intestinal transcript must be introduced by some form of post-transcriptional RNA editing mechanism.

Following the identification of RNA editing in the apolipoprotein B gene, it was subsequently discovered to occur in another mammalian gene which encodes a receptor for the excitatory amino acid glutamate and is expressed in neuronal cells. The editing of an A residue to a G residue in the transcript of this gene results in it encoding an arginine rather than the glutamine found in other related receptors and alters its properties so that it is permeable to calcium (Sommer *et al.*, 1991). The use of RNA editing in two different cases suggests that this novel mechanism in gene regulation may be widely used and that further cases will be identified in the future.

4.3   Regulation of RNA transport

The process of RNA splicing takes place within the nucleus, whereas the machinery for translating the spliced RNA is found in the cytoplasm. Hence the spliced mRNA must be transported to the cytoplasm if it is to direct protein synthesis. Such transport is known to occur through pores in the nuclear membrane and is an energy-dependent, active process and therefore potentially regulatable. The first example of regulation at this level has been described in the human immunodeficiency virus (HIV-1).

Figure 4.15 Regulation of HIV gene expression. Early in infection a low level of transcriptional initiation and elongation produces a small amount of the fully spliced viral RNA which encodes the viral Tat and Rev proteins. When these proteins are produced, the Tat protein enhances both the initiation and elongation of transcription, resulting in an increase in the production of the viral RNA. The Rev protein then acts post-transcriptionally to promote splicing/transport so the unspliced and singly spliced RNAs encoding the viral structural proteins (Gag, env) and the reverse transcriptase (pol) accumulate.

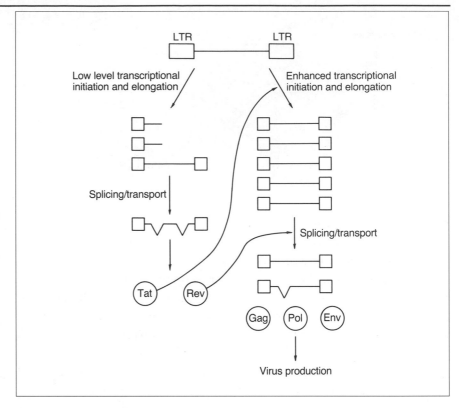

Thus early in infection of cells with HIV, the virus produces a high level of very short non-functional transcripts and a small amount of full length functional RNA (Section 3.3.2). The full length RNA is spliced with the removal of two introns so that the predominant transcript which appears in the cytoplasm is a fully spliced mRNA (Figure 4.15; for review, see Cullen, 1991). This transcript encodes the regulatory proteins Tat and Rev. As discussed in Chapter 3 (Section 3.3.2), the Tat protein acts to greatly increase the rate of transcription of the viral genome, as well as promoting transcriptional elongation so that predominantly full length RNAs are produced in the nucleus. However, at this second stage of infection most of the mRNAs that appear in the cytoplasm are either unspliced or have had only the first intron removed (Figure 4.15). As these transcripts encode the viral structural proteins, this allows the high-level production of viral particles which is necessary late in infection.

This change in the nature of the HIV RNA in the cytoplasm is dependent on the action of the Rev protein which is made early in infection (for review, see Cullen and Malim, 1991). Most interestingly, Rev does not affect the amounts of the different RNAs in the nucleus. Rather it is an RNA-binding protein and binds to a specific site (the Rev response

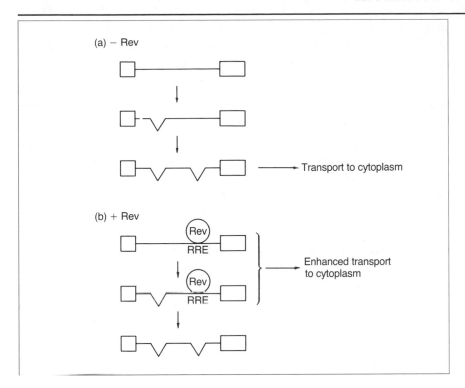

(a) − Rev

Transport to cytoplasm

(b) + Rev

Rev
RRE

Rev
RRE

Enhanced transport
to cytoplasm

Figure 4.16 Binding of Rev to its response element (RRE) in the HIV RNA promotes the transport of unspliced or singly spliced RNAs, which contain the binding site, at the expense of doubly spliced RNA, which lacks the binding site and therefore cannot bind Rev.

element) in the second intron of the HIV transcripts. This binding promotes the transport of these RNAs to the cytoplasm (Figure 4.16). As the fully spliced RNA lacks the Rev binding site, its transport is not accelerated and so the proportion of unspliced or singly spliced RNA in the cytoplasm increases. Hence Rev acts at the level of RNA transport, the first regulatory protein to do to. It is possible that Rev acts directly by interacting with a cellular transport system to promote the transport of RNA molecules to which it has bound. Alternatively it may act indirectly by displacing RNA splicing factors which otherwise prevent the RNA from being transported until it is fully spliced.

It has recently been shown that at least two other viruses, adenovirus and influenza virus, also encode proteins which can influence the transport of mRNAs from the nucleus to the cytoplasm (for review, see Krug, 1993). Although no similar case involving a cellular protein influencing RNA transport has yet been defined, it is unlikely that the regulation of RNA transport is confined to the eukaryotic viruses, and it seems probable that cases involving cellular genes will also be identified in the future. If such a mechanism does act on cellular RNAs, it would, for example, provide a convenient explanation of the existence of RNA species that are confined to the nucleus in specific tissues of the sea urchin (Section 4.1). Similarly, a mechanism which prevented the transport of unspliced RNA in genes regulated by processing versus discard decisions (Section 4.2) would represent a means of preventing wasteful translation of RNA species containing interruptions in the protein-coding sequence.

4.4 Regulation of RNA stability

4.4.1 Cases of regulation by alterations in RNA stability

Once the mRNA has entered the cytoplasm, the number of times that it is translated, and hence the amount of protein it produces, will be determined by its stability. The more rapidly degraded an RNA is, the less protein it will produce. Hence an effective means of gene regulation could be achieved by changing the stability of an RNA species in response to some regulating signal. A number of situations where the stability of a specific RNA species is changed in this way have been described (for reviews, see Peltz *et al.*, 1991; Sachs, 1993). Thus, for example, the mRNA for the milk protein, casein, turns over with a half life of around 1 h in untreated mammary gland cells. Following stimulation with the hormone prolactin, the half life increases to over 40 h (Guyette *et al.*, 1979), resulting in increased accumulation of casein mRNA and protein production in response to the hormone (Figure 4.17). Similarly, the increased production of the DNA-associated histone proteins in the S (DNA synthesis)

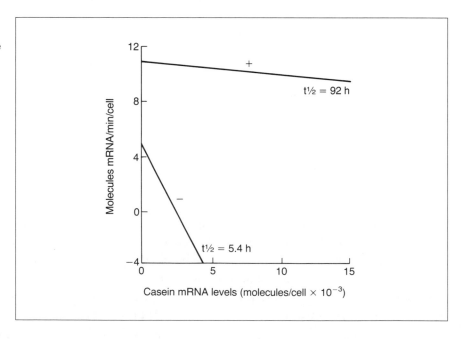

Figure 4.17 Difference in stability of the casein mRNA in the presence (+) or absence (−) of prolactin.

phase of the cell cycle is regulated in part by a 5-fold increase in histone mRNA stability that occurs in this phase of the cell cycle (for reviews of the regulation of histone gene expression, see Marzluff, 1992; Stein *et al.*, 1992). A representative selection of cases where mRNA stability is altered in a particular situation is given in Table 4.2.

4.4.2 Mechanisms of stability regulation

The first stage in defining the mechanism of changes in RNA stability is to identify the sequences within the RNA that are involved in mediating the observed alterations. This can be achieved by transferring parts of the gene encoding the RNA under study to another gene and observing the effect on the stability of the RNA expressed from the resulting hybrid gene. In a number of cases, short regions have been identified which can confer the pattern of stability regulation of the donor gene upon a recipient gene that is not normally regulated in this manner. In many cases such regions are located in the 3' untranslated region of the mRNA, downstream of the stop

Table 4.2 Regulation of RNA stability

mRNA	Cell type	Regulatory event	Increase or decrease in half life
Cellular oncogene c-*myc*	Friend erythroleukemia cells	Differentiation in response to dimethyl sulfoxide	Decrease from 35 to less than 10 min
c-*myc*	B cells	Interferon treatment	Decreased
c-*myc*	Chinese hamster lung fibroblasts	Growth stimulation	Increased
Epidermal growth factor receptor	Epidermal carcinoma cells	Epidermal growth factor	Increased
Casein	Mammary gland	Prolactin	Increased from 1 to 40 h
Vitellogenin	Liver	Estrogen	Increased 30-fold
Type 1 pro-collagen	Skin fibroblasts	Cortisol	Decreased
Type 1 pro-collagen	Skin fibroblasts	Transforming growth factor β	Increased
Histone	HeLa	Cessation of DNA synthesis	Decreased from 40 to 80 min
Tubulin	CHO	Accumulation of free tubulin subunits	Decreased 10-fold

codon that terminates production of the protein. Thus the cell-cycle-dependent regulation of histone H3 mRNA stability is controlled by a 30 nucleotide sequence at the extreme 3' end of the molecule. Similarly, the destabilization of the mRNA encoding the transferrin receptor in response to the presence of iron can be abolished by deletion of a 60 nucleotide sequence within the 3' untranslated region (for review, see Klausner *et al.*, 1993). Interestingly, both of these sequences have the potential to form stem–loop structures by intramolecular base pairing (Figure 4.18), suggesting that changes in stability might be brought about by alterations in the folding of this region of the RNA in response to a specific signal. A similar involvement of stem–loop structures in the 3' untranslated region has also been suggested to account for changes in the stability of chloroplast mRNAs during plant development (reviewed by Gruissem, 1989).

The localization of sequences involved in the regulated degradation of specific mRNA species to the 3' untranslated region is in agreement with the important role of this region in determining the differences in stability observed between different RNA species (Shaw and Kamen, 1986), suggesting that differences in RNA stability, whether between different RNA species or in a single RNA in different situations, may be controlled primarily by this region. Despite this, cases where other regions of the RNA

Figure 4.18 Similar stem–loop structures in the human ferritin and transferrin receptor mRNAs. Note the boxed conserved sequences in the unpaired loops and the absolute conservation of the boxed C residue, found within the stem, five bases 5' of the loop.

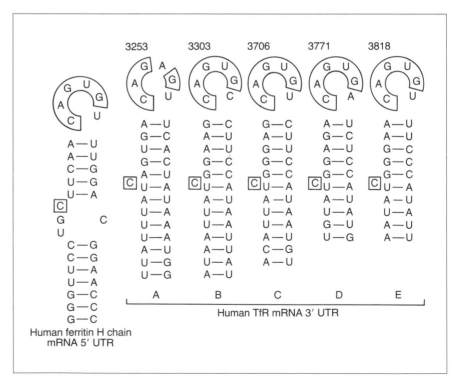

mediate the observed alterations in stability have also been described. The most extensively studied of such cases concerns the auto-regulation of the mRNA encoding the microtubule protein, β-tubulin, in response to free tubulin monomers (Pachter *et al.*, 1987; Yen *et al.*, 1988). This auto-regulation prevents the wasteful synthesis of tubulin when excess free tubulin, not polymerized into microtubules, is present and is caused by a destabilization of the tubulin mRNA. A short sequence, only 13 bases in length, from the 5' end of the β-tubulin mRNA is responsible for this destabilization and can confer the response on an unrelated mRNA. Most interestingly, these bases actually encode the first four amino acids of the tubulin protein, raising the possibility that the trigger for degradation of the tubulin mRNA might be the recognition of these amino acids in the tubulin protein, rather than the corresponding nucleotides in the tubulin RNA. In an elegant series of studies, Cleveland and colleagues (Yen *et al.*, 1988) showed that this was indeed the case. Thus changing the translational reading frame of this region, such that the identical nucleotide sequence encoded a different amino acid sequence, abolished the auto-regulatory response (Figure 4.19a), whereas changing the nucleotide sequence in a manner which did not alter the encoded amino acids (due to the degeneracy of the genetic code) left the response intact (Figure 4.19b).

This finding provides an explanation for the earlier observation that the tubulin RNA must be translated (at least in part) into protein for degradation to occur. Such a requirement for translation of the RNA to be

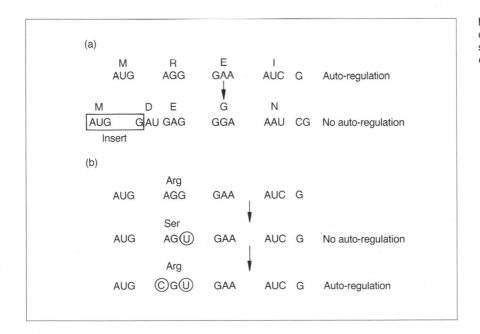

Figure 4.19 Effect of changes in the β-tubulin sequence on auto-regulation of tubulin mRNA stability.

degraded is also observed, however, in cases where the target sequence for degradation lies at the 3' end of the molecule, in a region that is not translated into protein. In the case of the histone mRNA, for example, introduction of stop codons resulting in premature termination of translation resulted in the abolition of the selective destabilization of the RNA following the cessation of DNA synthesis. Such an effect could also be achieved by inserting additional sequences between the normal termination codon and the previously discussed stem–loop structure which is involved in the regulation of degradation (Graves *et al.*, 1987; Figure 4.20).

These apparently contradictory observations can be combined into a model of RNA stability regulation, if it is assumed that the ribosome translating the RNA species carries with it a nuclease capable of degrading the RNA in response to a specific signal. In the case of tubulin this nuclease is triggered by a recognition event involving the first four amino acids emerging from the ribosome and then it degrades the RNA (Figure 4.21a). In contrast, in the case of histone mRNA the trigger is supplied by recognition of the 3' RNA sequence. However, translation of the upstream part of the molecule is still required in order to deliver the nuclease-bearing ribosome to the appropriate part of the molecule for recognition and degradation to occur (Figure 4.21b).

The finding that translation is also required both for the regulated degradation of other RNA species, such as the c-*myc* mRNA, as well as for the rapid turnover of short-lived mRNAs, such as that derived from another cellular oncogene, c-*fos*, suggests that this may represent a generally applicable mechanism controlling RNA degradation (for review, see Peltz and Jacobson, 1992).

4.4.3 Role of stability changes in regulation of gene expression

A consideration of the situations where changes in the stability of a particular RNA occur (Table 4.2) suggests that the majority have two features in

Figure 4.20 Effect of altering the relative position of the stop codon (x) and the stem–loop structure in histone mRNA on its degradation.

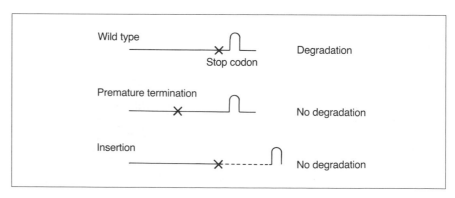

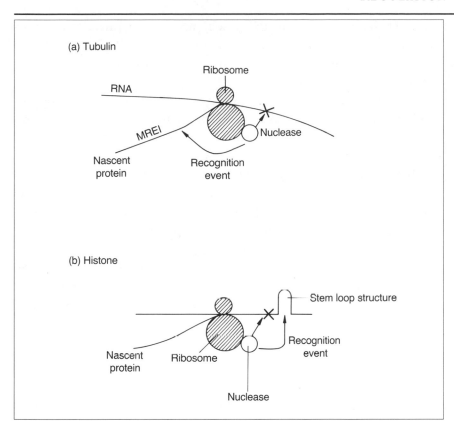

(a) Tubulin

Ribosome

RNA

MREI

Nascent
protein

Nuclease

Recognition
event

(b) Histone

Stem loop structure

Nascent
protein

Ribosome

Recognition
event

Nuclease

Figure 4.21 Putative mechanism for recognition of degradation signals in nascent tubulin protein (a) or histone mRNA (b) by a ribosome-associated nuclease.

common. First, changes in stability of a particular mRNA are very often accompanied by parallel alterations in the transcription rate of the corresponding gene. Thus prolactin treatment of mammary gland cells results in a 2- to 4-fold increase in casein gene transcription and the increased stability of histone mRNA in the S phase of the cell cycle is accompanied by a 3- to 5-fold increase in transcription of the histone genes. Secondly, cases where RNA stability is regulated are very often those where a rapid and transient change in the synthesis of a particular protein is required. Thus synthesis of the histone proteins is necessary only at one particular phase of the cell cycle, when DNA is being synthesized. Following cessation of DNA synthesis a rapid shut-off in the synthesis of unnecessary histone proteins is required. Similarly, following the cessation of hormonal stimulation it would be highly wasteful to continue the synthesis of hormonally dependent proteins such as casein or vitellogenin. In the case of the cellular oncogene c-*myc*, whose RNA stability is transiently increased when cells are stimulated to grow, such continued synthesis would not only be highly wasteful but is also potentially dangerous to the cell. Thus this growth regulatory protein is only required for a short period when cells are entering the growth phase, its continued inappropriate synthesis at other times carrying the risk of disrupting cellular growth regulatory mechanisms, possibly resulting in transition to a cancerous state (Chapter 8).

These considerations suggest that alterations in RNA stability are used as a significant supplement to transcriptional control in cases where rapid changes in the synthesis of a particular protein are required. Thus, if transcription is shut off in response to withdrawal of a particular signal, inappropriate and metabolically expensive protein synthesis will continue for some time from pre-existing mRNA, unless that RNA is degraded rapidly. Similarly, rapid onset of the expression of a particular gene can be achieved by having a relatively high basal level of transcription with high RNA turnover in the absence of stimulation, allowing rapid onset of translation from pre-existing RNA following stimulation. Hence, as with alternative RNA processing, cases where RNA stability is regulated represent an adaptation to the requirements of a particular situation and do not affect the conclusion that regulation of gene expression occurs primarily at the level of transcription.

4.5   Regulation of translation

4.5.1   Cases of translational control

The final stage in the expression of a gene is the translation of its mRNA into protein. In theory, therefore, the regulation of gene expression could be achieved by producing all possible mRNA species in every cell and selecting which were translated into protein in each individual cell type. The evidence that different cell types have very different cytoplasmic RNA populations (Chapter 1) indicates, however, that this extreme model is incorrect. None the less, the regulation of translation, such that a particular mRNA is translated into protein in one situation and not another, does occur in some special cases (for review, see Altmann and Trachsel, 1993).

The most prominent of such cases is that of fertilization. In the unfertilized egg protein synthesis is slow, but upon fertilization of the egg by a sperm a tremendous increase in the rate of protein synthesis occurs. This increase does not require the production of new mRNAs after fertilization. Rather, it is mediated by pre-existing maternal RNAs which are present in the unfertilized egg but are only translated after fertilization. Although in many species such translational control produces only quantitative changes in protein synthesis, in others it can affect the nature as well as the quantity of the proteins being made before and after fertilization. Thus in

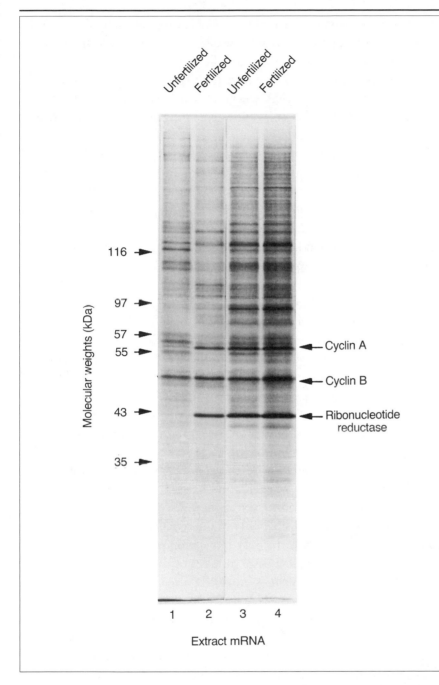

Figure 4.22 Translational control in the clam *Spisula solidissima*. Different proteins are synthesized *in vivo* before fertilization (track 1) and after fertilization (track 2). If, however, RNA is isolated either before (track 3) or after fertilization (track 4) and translated *in vitro* in a cell-free system, identical patterns of proteins are produced. Hence the difference in the proteins produced *in vivo* from identical RNA populations must be due to translational control.

the clam *Spisula solidissima* some new proteins appear after fertilization, while others which are synthesized in large amounts before fertilization are repressed thereafter. However, the RNA populations present before and after fertilization are identical (Figure 4.22), indicating that translational control processes are operating (for review, see Standart, 1992).

Similar translational control processes operating on many RNA species also occur following infection with the large DNA viruses, such as adenovirus or herpes simplex virus, and following the exposure of cells to elevated temperatures (heat shock). In both these cases, the translation of most pre-existing cellular RNA species is repressed, while the translation of either the viral mRNAs or of those encoding the heat-shock proteins (Chapter 3) occurs at high levels.

As well as producing parallel changes in the translation of many RNA species, translational control processes may also operate on individual RNAs in a particular cell type. Thus, for example, the rate of translation of the globin RNA in reticulocytes is regulated in response to the availability of the heme co-factor which is required for the production of hemoglobin. Similarly, the translation of the RNA encoding the iron-binding protein, ferritin, is regulated in response to the availability of iron.

4.5.2   Mechanism of translational control

In principle, translational regulation could operate via modifications in the cellular translational apparatus affecting the efficiency of translation of particular RNAs or by modifications in the RNA itself which affect the way in which it is translated by the ribosome. Evidence is available indicating that both these types of mechanism are used in different cases.

Thus, in the absence of heme a cellular protein kinase in the reticulocyte becomes active and phosphorylates the protein initiation factor eIF2, resulting in its inactivation. Since this factor is required for the initiation of protein synthesis, translation of the globin RNA ceases until heme is available. However, the use of such a mechanism, in which total inactivation of the cellular translational apparatus is used to regulate the translation of a single RNA, is possible only in the reticulocyte, where the globin protein constitutes virtually the only translation product.

In other cell types, where a large number of different RNA species are expressed, such a mechanism is normally used only where large-scale repression of many different RNA species occurs. Thus phosphorylation of eIF2 (produced by a different kinase to that activated by the absence of heme) also occurs following infection with viruses, such as adenovirus, which inhibit the translation of most cellular mRNAs, and it is probable that a similar mechanism is also responsible for the repression of cellular protein synthesis following heat shock.

Most other situations where the translation of only a single RNA species is regulated, however, are likely to operate through sequences within the specific RNA itself. Indeed, the preferential translation of the RNAs encoding the heat-shock protein, which accompanies the repression of

Table 4.3 Regulation of the transferrin receptor and ferritin genes

	Effect of iron on protein production	Mechanism	Position of stem–loop structure
Ferritin	Increased	Increased mRNA translation	5' untranslated region
Transferrin receptor	Decreased	Decreased mRNA stability	3' untranslated region

most cellular protein synthesis following heat shock, is mediated by short sequences contained within the 5' untranslated regions of these RNAs upstream of the point at which protein synthesis is initiated (Hultmark *et al.*, 1986). Presumably, these sequences allow the modified translational apparatus that exists after heat shock to recognize and translate these mRNAs.

Several other cases where translational regulation of particular mRNA species is mediated by sequences in their 5' untranslated region have been defined. Thus, the enhanced translation of the ferritin mRNA in response to iron is mediated by a sequence in this region which can fold into a stem–loop structure (for review, see Klausner *et al.*, 1993). The structure of this stem–loop is very similar to that found in the 3' untranslated region of the transferrin receptor mRNA, whose stability is negatively regulated by the presence of iron (Figure 4.18 and Table 4.3). This has led to the suggestion that such loops may represent functionally equivalent iron response elements whose opposite effects on gene expression are dependent upon their position (5' or 3') within the RNA molecule. This idea was confirmed by transferring the transferrin receptor stem–loop to the 5' end of an unrelated RNA, resulting in the iron-dependent enhancement of its translation. The identical structure is therefore capable of mediating opposite effects on RNA stability and translation, depending on its position within the RNA molecule. Such an apparent paradox can be explained if it is assumed that, in response to the presence of iron, the stem–loop structure is an iron response element (IRE) which unfolds in the presence of iron (Figure 4.23). In the ferritin mRNA, where the element is in the 5' end, this will allow unimpeded movement of the ribosome along the mRNA and increased translation. In contrast, in the transferrin receptor mRNA, where this element is in the 3' untranslated region, such unfolding presumably renders the RNA susceptible to nuclease degradation at an increased rate. In agreement with this model, an IRE-binding protein (IRE-BP) has been identified which binds to the stem–loop element in both the ferritin and transferrin receptor mRNAs.

Figure 4.23 Role of iron-induced unfolding of the stem–loop structure in producing increased translation of the ferritin mRNA (panel a) and increased degradation of the transferrin receptor mRNA (panel b). In each case the unfolding of the stem loop is dependent on the dissociation of a binding protein (IRE-BP) whose ability to bind to the stem–loop decreases dramatically in the presence of iron.

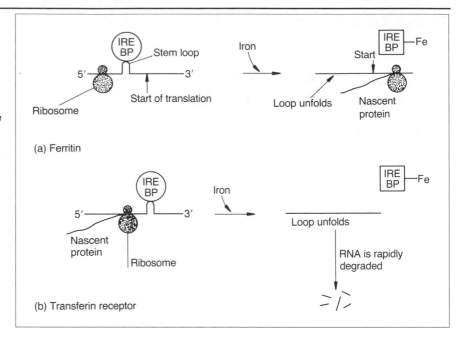

(a) Ferritin

(b) Transferin receptor

The RNA-binding activity of IRE-BP increases dramatically in cells that have been deprived of iron, suggesting that its binding normally stabilizes the stem–loop structure (for review, see Klausner *et al.*, 1993).

Not all cases of translational regulation mediated by sequences in the 5' untranslated region operate via such stem–loop structures, however. Increased expression of the yeast regulatory protein GCN4 in response to amino acid starvation is caused by increased translation of its RNA (for review, see Altmann and Trachsel, 1993). The translational regulation of this molecule is mediated by short sequences within the 5' untranslated

Figure 4.24 Presence of short open reading frames capable of producing small peptides in the 5' untranslated region of the yeast GCN4 RNA. Translation of the RNA to produce these small proteins supresses translation of the GCN4 protein. The position of the methionine residue beginning each of the small peptides is indicated together with the number of additional amino acids incorporated before a stop codon is reached. When high amino acid levels are present, the small proteins are made and the production of GCN4 is therefore suppressed. When amino acid levels fall, the eIF-2 translational initiation factor is phosphorylated. This prevents the translation of three of the four short peptides and thereby promotes GCN4 production.

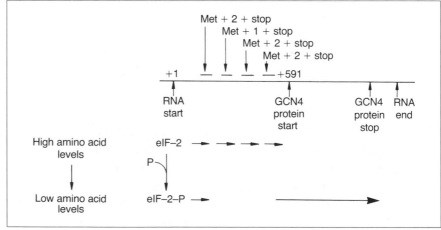

region of the RNA, upstream of the start point for translation of the GCN4 protein. Most interestingly, such sequences are capable of being translated to produce peptides of two or three amino acids (Figure 4.24). Following translational initiation at the second, third or fourth of these upstream sequences to yield such small peptides, the ribosome apparently fails to reinitiate at the translational start point for GCN4 production and hence this protein is not synthesized. Following amino acid starvation, the production of these small peptides is suppressed and production of GCN4 correspondingly enhanced. This switch in initiation also involves the phosphorylation of the eIF2 factor which is involved in translational regulation in response to heme or viral infection (see above). Thus following amino acid starvation, transfer RNA molecules lacking a bound amino acid accumulate and this activates an enzyme which phosphorylates eIF2. The decreased activity of eIF2 caused by its phosphorylation then results in the ribosome not initiating at three of the upstream sites in the 5' untranslated region which allows increased initiation at the downstream site, leading to enhanced GCN4 production.

Interestingly, alternative initiation codons are also found in the gene encoding a mammalian liver regulatory protein, as discussed in Chapter 7 (Section 7.4.2). However, in this case the two initiation codons are used to produce two different forms of the protein. One of these contains an N-terminal sequence which allows it to activate transcription whereas the other form does not and so lacks this ability and can therefore interfere with the stimulatory activity of the activating form. Such alternative translation of the same mRNA thus parallels the use of alternative splicing to produce different protein products from the same gene.

Study of the cases of translational control where the mechanisms have been defined thus makes it clear that although sequences in the 5' untranslated region of particular mRNAs are involved in the translational regulation of their expression, the mechanism by which they do so may differ dramatically in different cases.

Although the 5' untranslated region is an obvious location for sequences involved in mediating translational control, cases where sequences in the 3' untranslated region play a role in the regulation of translation have been reported. Thus, sequences in this region are involved in modulating the efficiency of translation of the specific mRNAs which occurs upon fertilization of the egg (Section 4.5.1) as well as in mediating the increased translation of the mRNA encoding tissue plasminogen activator (t-PA), which occurs during meiotic maturation of mouse oocytes. In this latter case these sequences appear to act by affecting the addition of adenosine residues to the 3' end of the RNA, such polyadenylation being necessary for the translation of the RNA (for reviews, see Jackson, 1993; Wormington, 1993). Interestingly, sequences in the same region of the t-PA mRNA are also involved in causing the subsequent destabilization of the mRNA and its rapid degradation later in the process of oocyte maturation, further emphasizing the relationship between the various forms of post-transcriptional control.

Other cases have been described in which sequences in the 3' untranslated region regulate translation directly rather than via polyadenylation. Thus a 60 bp sequence in the 3' untranslated region of the protamine mRNAs mediates their increased translation during spermatogenesis, with this increased translation being observed at a time when no change in the polyadenylation of these RNAs is observed (Braun, 1991). Although most cases of this type are likely to involve specific regulatory proteins binding to the 3' untranslated region and controlling translation of the mRNA, one case has recently been described in which regulation is dependent on a regulatory RNA. Thus in the nematode the translation of the mRNA derived from the *lin*-14 gene is inhibited by a small RNA derived from the *lin*-4 gene which appears to bind to the 3' untranslated region of the *lin*-14 mRNA and block its translation (Wickens and Takayama, 1994). Hence as in the case of translational modulation by sequences at the 5' end of the mRNA, multiple mechanisms mediate the effects of sequences at the 3' end of the mRNA on translational efficiency.

In summary, therefore, it is clear that cases of translational control can be mediated by sequences in various parts of the RNA and can involve secondary structure, use of different translation initiation codons, or the regulation of polyadenylation.

4.5.3 Significance of translational control

Many of the cases of translational control occur in situations where very rapid responses are required. Thus, following fertilization a very rapid activation of cellular growth processes is required. Similarly, following heat shock it is necessary to shut down rapidly the synthesis of most enzymes and structural proteins and begin to synthesize the protective heat-shock proteins. Such regulation can be achieved rapidly by translational control, supplemented by increased transcription of the genes encoding the heat-shock proteins. Once again, as with other cases of post-transcriptional regulation, translational control can be viewed as supplementing the regulation of transcription in order to meet the requirements of particular specialized cases. In the case of the heat-shock proteins, this combination of transcriptional and translational control is further supplemented by an increased stability of the heat-shock mRNAs following exposure to elevated temperature, providing a further means of producing a rapid and effective response (Theodorakis and Morimoto, 1987).

The case of the yeast GCN4 protein provides a different aspect to translational control, however. This protein is a transcriptional regulator, which increases the transcription of several genes encoding the enzymes of amino acid biosynthesis in response to a lack of one or more amino acids. In this case the synthesis of a transcriptional regulatory protein is regulated by translational control. When taken together with the increasing evidence (Chapter 7) that many mammalian transcriptional regulatory molecules may pre-exist in an inactive form and be activated by protein modifications (e.g. phosphorylation), this suggests that it may be necessary for the cell to control the expression of some of its transcriptional regulatory molecules at levels other than that of gene transcription. Translational control may be one means by which such regulation is achieved.

4.6 Conclusions

A wide variety of cases exist in which gene expression can be regulated at levels other than transcription. In some lower organisms such post-transcriptional regulation may constitute the predominant form of gene control. In mammals, however, it appears to represent an adaptation to particular situations, including the need to generate two closely related proteins from one gene by alternative RNA splicing, or to respond rapidly to the withdrawal of hormonal stimulation or stress by the regulation of RNA stability or translation. It is likely that many more cases of post-transcriptional regulation will be described in the future, whilst this form of regulation may be the predominant one for the proteins which themselves regulate the transcription of other genes.

References

Altmann, M. and H. Trachsel 1993. Regulation of translational initiation and modulation of cellular physiology. *Trends in Biochemical Sciences* **18**, 429–32.

Baker, B. S. 1989. Sex in flies: the splice of life. *Nature* **340**, 521–4.

Bandziulis, R. J., M. S. Swanson and G. Dreyfuss 1989. RNA-binding proteins as developmental regulators. *Genes and Development* **3**, 431–7.

Bingham, P. M., T.-B. Chou, I. Mims and Z. Zachar 1988. On/off regulation of gene expression at the level of gene splicing. *Trends in Genetics* **4**, 134–8.

Braun, R. E. 1991. Temporal translational regulation of the protamine 1 gene during spermatogenesis. *Enzyme* **44**, 120–8.

Breitbart, R. E., A. Andreadis and B. Nadal-Ginard 1987. Alternative splicing: a ubiquitous mechanism for the generation of multiple protein isoforms from different genes. *Annual Review of Biochemistry* **56**, 467–95.

Cullen, B. R. 1991. Regulation of HIV-1 gene expression. *FASEB Journal* **5**, 2361–8.

Cullen, B. R. and H. Malim 1991. The HIV-1 Rev protein: prototype of a novel class of eukaryotic transcriptional regulators. *Trends in Biochemical Sciences* **16**, 341–8.

Danner, D. and P. Leder 1985. Role of an RNA cleavage/poly (A) addition site in the production of membrane bound and secreted IgM mRNA. *Proceedings of the National Academy of Sciences of the USA* **82**, 8658–62.

Davies, K. 1992. Imprinting and splicing join together. *Nature* **360**, 492.

Delsert, C. D. and M. G. Rosenfeld 1992. A tissue specific small nuclear ribonucleoprotein and the regulated splicing of the calcitonin/calcitonin gene related protein transcript. *Journal of Biological Chemistry* **267**, 14573–9.

Graves, R. A., N. B. Pandey, N. Chodchoy and W. F. Marzluff 1987. Translation is required for regulation of histone mRNA degradation. *Cell* **48**, 615–26.

Gruissem, W. 1989. Chloroplast gene expression: how plants turn their plastids on. *Cell* **56**, 161–70.

Guyette, W. A., R. A. Matusik and J. M. Rosen 1979. Prolactin-mediated transcriptional and post-transcriptional control of casein gene expression. *Cell* **17**, 1013–23.

Hodges, P. and J. Scott 1992. Apolipoprotein B mRNA editing: a new tier for the control of gene expression. *Trends in Biochemical Sciences* **17**, 77–81.

Hultmark, D., R. Klemenz and W. Gehring 1986. Translational and transcriptional control elements in the untranslated leader of the heat-shock gene hsp 22. *Cell* **44**, 429–38.

Jackson, R. J. 1993. Cytoplasmic regulation of mRNA function. *Cell* **74**, 9–14.

Klausner, R. D., T. A. Rouault and J. B. Harford 1993. Regulating the fate of cellular iron metabolism. *Cell* **72**, 19–28.

Krainer, A. R., G. C. Conway and D. Kozak 1990. The essential pre-mRNA splicing factor SF2 influences 5' splice site selection by activating proximal sites. *Cell* **62**, 35–42.

Krug, R. M. 1993. The regulation of mRNA transport from nucleus to cytoplasm. *Current Opinion in Cell Biology* **5**, 944–9.

Lamond, A. I. 1991a. Nuclear RNA processing. *Current Opinion in Cell Biology* **3**, 493–500.

Lamond, A. I. 1991b. ASF/SF2 a splice selector. *Trends in Biochemical Sciences* **16**, 452–3.

Latchman, D. S. 1990. Cell-type specific splicing factors and the regulation of alternative mRNA splicing. *New Biologist* **2**, 297–303.

Leff, S. E., M. G. Rosenfeld and R. M. Evans 1986. Complex transcriptional units: diversity in gene expression by alternative RNA processing. *Annual Review of Biochemistry* **55**, 1091–117.

Leff, S. E., R. M. Evans and M. G. Rosenfeld 1988. Splice commitment dictates neuron-specific alternative RNA processing in calcitonin-CGRP gene expression. *Cell* **48**, 517–24.

Lund, E., B. Kahan and J. E. Dahlberg 1985. Differential control of U1 small nuclear RNA expression during mouse development. *Science* **229**, 1271–4.

McKeown, M. 1992. Alternative mRNA splicing. *Annual Review of Cell Biology* **8**, 133–55.

Maniatis, T. and R. Reed, 1987. The role of small ribonucleoprotein particles in pre-mRNA splicing. *Nature* **325**, 673–8.

Marzluff, W. F. 1992. Histone 3' ends: essential and regulatory function. *Gene Expression* **2**, 93–7.

Mayeda, A. and A. R. Krainer 1992. Regulation of alternative pre-mRNA splicing by hnRNPA1 and splicing factor SF2. *Cell* **68**, 365–75.

Pachter, J. S., T. J. Yen and D. W. Cleveland 1987. Auto regulation of tubulin expression is achieved through degradation of polysomal tubulin mRNAs. *Cell* **51**, 283–92.

Peltz, S. W. and A. Jacobson 1992. mRNA stability in *trans*-it. *Current Opinion in Cell Biology* **4**, 979–83.

Peltz, S. W., G. Brewer, P. Bernstein and J. Ross 1991. Regulation of mRNA turnover in eukaryotic cells. *Critical Reviews in Eukaryotic Gene Expression* **1**, 99–126.

Powell, D. J., J. M. Freidman, A. J. Oulette, K. S. Krauter and J. E. Darnell, Jr 1984. Transcriptional and post-transcriptional control of specific messenger RNAs in adult and embryonic liver. *Journal of Molecular Biology* **179**, 21–35.

Rio, D. S. 1991. Regulation of *Drosophila* P element transposition. *Trends in Genetics* **7**, 282–87.

Rosenfeld, M. G., S. G. Amara and P. M. Evans 1984. Alternative RNA processing: determining neuronal phenotype. *Science* **225**, 1315–20.

Sachs, A. B. 1993. Messenger RNA degradation in eukaryotes. *Cell* **74**, 413–21.

Sharp, P.A. 1987. Splicing of messenger RNA precursors. *Science* **235**, 766–71.

Shaw, G. and R. Kamen 1986. A conserved AU sequence from the 3' untranslated region of GM-CSF mRNA mediates selective mRNA degradation. *Cell* **46**, 659–67.

Sommer, B., M. Kohler, R. Sprengel and P. H. Seeburg 1991. RNA editing in brain controls a determinant of ion flow in glutamate gated channels. *Cell* **67**, 11–19.

Standart, N. M. 1992. Masking and unmasking of maternal mRNA. *Seminars in Developmental Biology* **3**, 367–79.

Stein G. S., J. L. Stein, A. J. van Wijnen and J. B. Liam 1992. Regulation of histone gene expression. *Current Opinion in Cell Biology* **4**, 166–73.

Theodorakis, N. G. and R. I. Morimoto 1987. Post-transcriptional regulation of hsp70 in human cells: effects of heat shock, inhibition of protein synthesis and adenovirus infection on translation and mRNA stability. *Molecular and Cellular Biology* 7, 4357–68.

Valcarcel, J., R. Singh, P. D. Zamore and M. R. Green 1993. The protein sex-lethal antagonizes the splicing factor U2AF to regulate splicing of transformer pre-mRNA. *Nature* 362, 171–5.

Wickens, M. and K. Takayama 1994. RNA: deviants or emissaries. *Nature* 367, 17–18.

Wold, B. J., W. H. Klein, B. R. Hough-Evans, R. J. Britten and E. H. Davidson 1978. Sea urchin embryo mRNA sequences expressed in the nuclear RNA of adult tissues. *Cell* 14, 941–50.

Wormington, M. 1993. Poly(A) and translation: developmental control. *Current Opinion in Cell Biology* 5, 950–4.

Yeakley, J. M., F. Hedjrun, J.-P. Morfin, N. Merillat, M. G. Rosenfeld and R. B. Emson 1993. Control of calcitonin/calcitonin gene-related peptide pre-mRNA processing by constitutive intron and exon elements. *Molecular and Cellular Biology* 13, 5999–6011.

Yen, T. J., P. S. Machlin and D. W. Cleveland 1988. Autoregulated instability of β-tubulin by recognition of the nascent amino-acids of β-tubulin. *Nature* 334, 580–5.

Young, R. A., O. Hagenbuchle and U. Schibler 1981. A single mouse α-amylase gene specifies two different tissue specific mRNAs. *Cell* 23, 451–8.

Transcriptional control – chromatin structure

5

5.1 Introduction

Having established that the primary control of eukaryotic gene expression lies at the level of transcription, it is necessary to investigate the mechanisms responsible for this effect. The fact that regulation at transcription is also responsible for the control of gene expression in bacteria suggests that insights into these procedures obtained in these much simpler organisms (reviewed by Gottesman, 1984; Travers, 1993) may be applicable to higher organisms. It is possible therefore that regulation of transcription in eukaryotes might occur by means of a protein, present in all tissues, which binds to the promoter region of a particular gene and prevents its expression. In one particular cell type, or in response to a particular signal such as heat shock, this protein would be inactivated either directly (Figure 5.1a) or by binding of another factor (Figure 5.lb) and would no longer bind to the gene. Hence transcription would occur only in the one cell type or in response to the signal. This mechanism is based on that regulating the expression of the *lac* operon containing the genes encoding proteins required for the metabolism of lactose. This operon is normally repressed by the *lac* repressor protein; binding of lactose to this protein, however, results in its inactivation and allows transcription of the operon (reviewed by Miller and Reznikoff, 1980).

Alternatively, the fact that most eukaryotic genes are inactive in most tissues, and become active only in one tissue or in response to a particular signal, suggests that it may be more economical to have a system in which the gene is constitutively inactive in most tissues, without any repressor being required. Activation of the gene would then require a particular factor binding to its promoter. The specific expression pattern of the gene would be controlled by the presence of this factor only in the expressing cell type (Figure 5.2a) or, alternatively, by the factor's requirement for a co-

Figure 5.1 Model for the activation of gene expression in a particular tissue by inactivation (a) or binding out (b) of a repressor present in all tissues.

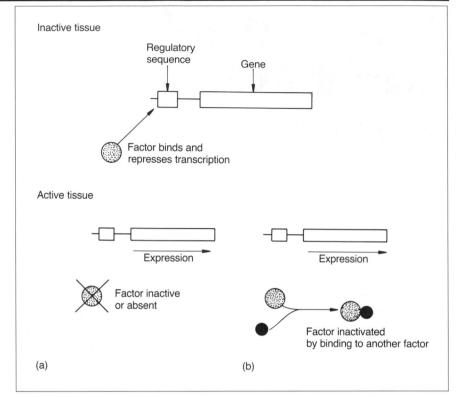

factor such as a steroid hormone to convert it to an active form (Figure 5.2b). This mechanism is based on the regulation of the arabinose operon in *Escherichia coli* in which binding of the substrate arabinose to the regulatory araC protein allows it to induce transcription of the genes required for the metabolism of arabinose (reviewed by Raibaud and Schwartz, 1984).

Hence, based on the known mechanisms of gene regulation in bacteria, it is possible to produce models of how gene regulation might operate in eukaryotes. Indeed, a considerable amount of evidence indicates that many cases of gene regulation do use the activation-type mechanisms seen in the arabinose operon. Thus, as will be discussed in Chapters 6 and 7, the effects of glucocorticoid and other steroid hormones on gene expression are mediated by the binding of the steroid to a receptor protein. This activated complex then binds to particular sequences upstream of steroid-responsive genes and activates their transcription (reviewed by Beato, 1989).

Even in the case of steroid hormones, however, such mechanisms cannot account entirely for the regulation of gene expression. In the chicken, administration of the steroid hormone estrogen results in the transcrip-

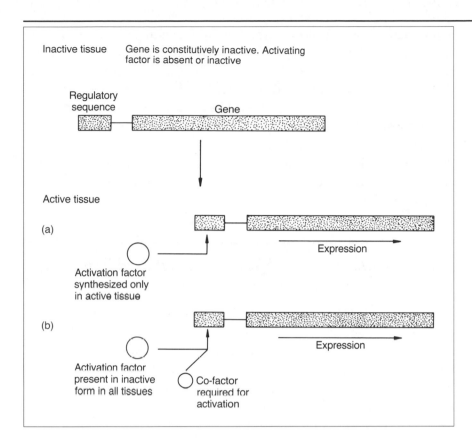

Inactive tissue   Gene is constitutively inactive. Activating
                     factor is absent or inactive

Regulatory
sequence   Gene

Active tissue

(a)

                                            Expression

Activation factor
synthesized only
in active tissue

(b)

                                            Expression

Activation factor
present in inactive   Co-factor
form in all tissues   required for
                       activation

Figure 5.2 Model for the activation of gene expression in a particular tissue by an activator present only in that tissue (a) or activated by a co-factor (b) only in that tissue.

tional activation in oviduct tissue of the gene encoding ovalbumin, as discussed in Chapter 3. In the liver of the same organism, however, estrogen treatment has no effect on the ovalbumin gene but instead results in the activation of a completely different gene, encoding the protein vitellogenin. Such tissue-specific differences in the response to a particular treatment are, of course, entirely absent in single-celled bacteria and cannot be explained simply on the basis of the activation of a single DNA-binding protein by the hormone. Similarly, models of this type cannot explain the data obtained by Becker *et al.* (1987) in their studies of the regulation of the rat tyrosine aminotransferase gene. All the protein factors binding to the regulatory regions of this gene were detectable in both the liver, where the gene is expressed, and in other tissues, where no expression is observed. Moreover, the proteins isolated from these tissues were equally active in binding to the appropriate region of the gene in deproteinized DNA. Only in the liver, however, was actual binding of these factors to the DNA of the gene within its normal chromosomal structure detectable. Once again, these data cannot be explained solely on the basis of models in which proteins stimulate or inhibit gene expression by binding to appropriate DNA sequences. Rather, an understanding of eukaryotic gene regulation will require a knowledge of the ways in which the relatively short-term regulatory processes mediated by the binding of proteins to specific DNA sequences in this manner interact with the much longer-term regulatory processes, which establish and maintain the differences between particular tissues and also control their response to treatment with effectors such as steroids. These long-term control processes will be discussed in this chapter, and the DNA sequences and proteins which actually regulate transcription in response to particular signals will be discussed in Chapters 6 and 7.

5.2   Commitment to the differentiated state and its stability

It is evident that the existence of different tissues and cell types in higher eukaryotes requires mechanisms that establish and maintain such differences, and that these mechanisms must be stable in the long term. Thus, despite some exceptions (Section 2.2.3), tissues or cells of one type do not, in general, change spontaneously into another cell type. Inspection of mammalian brain tissue does not reveal the presence of cells typical of liver or kidney, and antibody-producing B cells do not spontaneously change into muscle cells when cultured in the laboratory. Hence, cells must be capable of maintaining their differentiated state, either within a tissue or during prolonged periods in culture.

Indeed, the long-term control processes that achieve this effect must regulate not only the ability to maintain the differentiated state more or less indefinitely but also the observed ability of cells to remember their particular cell type, even under conditions when they cannot express the characteristic features of that cell type. Thus in the experiments of Coon (1966) it was possible to regulate the behavior of cartilage-producing cells in culture according to the medium in which they were placed. In one medium the cells were capable of expressing their differentiated phenotype and produced cartilage-forming colonies synthesizing an extracellular matrix containing chondroitin sulfate. By contrast, in a medium favoring rapid division the cells did not form such colonies and, instead, divided rapidly and lost all the specific characteristics of cartilage cells, becoming indistinguishable from undifferentiated fibroblast-like cells in appearance. None the less, even after 20 generations in this rapid growth medium the cells could resume the appearance of cartilage cells and synthesis of chondroitin sulfate if returned to the appropriate medium. This process did not occur if other cell types or undifferentiated fibroblast cells were placed in an identical medium. Hence the cartilage cells were capable not only of maintaining their differentiated cell type in a particular medium supplying appropriate signals, but also of remembering it in the absence of such signals and returning to the correct differentiated state when placed in the appropriate medium.

These experiments lead to the idea that cells belong to a particular lineage and that mechanisms exist to maintain the commitment of cells to a particular lineage, even in the absence of the characteristics of the differentiated phenotype. A variety of evidence exists to suggest that cells become committed to a particular differentiated stage or lineage well before they express any features characteristic of that lineage and that such commitment can be maintained through many cell generations. Perhaps the most dramatic example of this effect occurs in the fruit fly, *Drosophila melanogaster*. In this organism, the larva contains many disks consisting of undifferentiated cells located at intervals along the length of the body and indistinguishable from each other in appearance. Eventually, these imaginal disks (Hadorn, 1968) will form the structures of the adult, the most anterior pair producing the antennae and others producing the wings, legs, etc. In order to do this, however, the disks must pass through the intermediate pupal state where they receive the appropriate signals to differentiate into the adult structures. If they are removed from the larva and placed directly in the body cavity of the adult, they will remain in an undifferentiated state, since the signals inducing differentiation will not be received (Figure 5.3). This process can be continued for many generations. Thus the disk removed from the adult can be split in two, one half being used to propagate the cells by being placed directly in another adult and the other half being tested to see what it will produce when placed in a larva that is allowed to proceed to an adult through the pupal state. When this is done, it is found that a disk which would have given rise to an antenna can still do so when placed in a larva after many generations of passage through adult organisms in an undifferentiated state. This is the case even if the disk is placed at a position within the larva very different from that normally occupied by the imaginal disk producing the antenna. The cells of the imaginal disk within the normal larva must therefore have undergone a commitment event to the eventual production of a particular differentiated state, which they can maintain for many generations, prior to having ever expressed any features characteristic of that differentiated state. These examples of the stability of the differentiated state and commitment to it imply the existence of long-term regulatory processes capable of maintaining such stability. However, the existence of cases

Figure 5.3 The use of larval imaginal disks to demonstrate the stability of the committed state. The disk cells maintain their commitment to produce a specific adult structure even after prolonged growth in an undifferentiated state in successive adults.

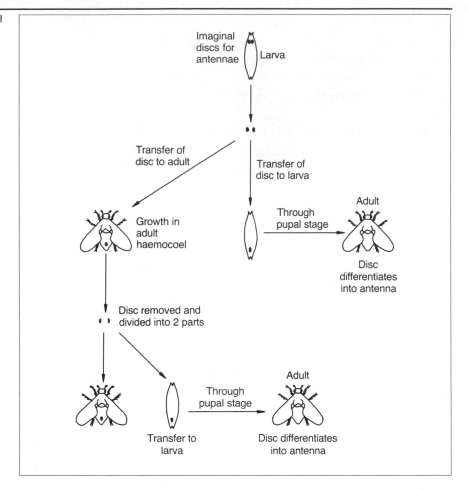

where such stability breaks down, either artificially following nuclear transplantation or more naturally as in lens regeneration (Section 2.2.3), indicates that such processes, although stable, cannot be irreversible. Indeed, even in the imaginal disks of *Drosophila* the stability of the committed state can be broken down by culture for long periods in adults, the disks giving rise eventually to tissues other than the one intended originally when placed in the larva. This breakdown of the committed state is not a random process but proceeds in a highly characteristic manner. Wing cells are the first abnormal cells produced by a disk that should produce an eye, with other cell types being produced only subsequently. Similarly, a disk that should produce a leg never produces genitalia, although it can produce other cell types. The various transitions that can occur in this system are summarized in Figure 5.4.

Most interestingly, these various transitions are precisely those observed in the homeotic mutations in *Drosophila* (Section 7.2.2) which convert

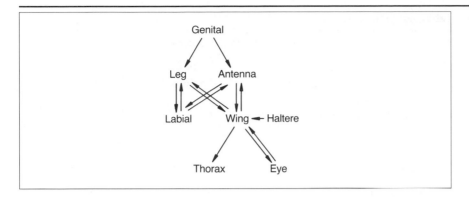

one adult body part into another. Hence each of these transitions is likely to be controlled by a specific gene whose activity changes when commitment breaks down.

These examples and the other experiments discussed in Chapter 2 (Section 2.2) eliminate irreversible mechanisms such as DNA loss as a means of explaining the process of commitment to the differentiated state. It is thought, therefore, that the semi-stability of this process and its propagation through many cell generations is due to the establishment of a particular pattern of association of the DNA with specific proteins. The structure formed by DNA and its associated proteins is known as chromatin. An understanding of how long-term gene regulation is achieved therefore requires a knowledge of the structure of chromatin.

5.3   Chromatin structure

If the DNA in a single human individual were to exist as an extended linear molecule, it would have a length of 5×10^{10}km and would extend 100 times the distance from the Earth to the Sun (Lewin, 1980). Clearly this is not the case. Rather, the DNA is compacted by folding in a complex with specific nuclear proteins into the structure known as chromatin (for review, see Igo-Kemenes *et al.*, 1982). Of central importance in this process are the five types of histone proteins (Table 5.1) whose high proportion of positively charged amino acids neutralizes the net negative charge on the DNA and allows folding to occur. The basic unit of this folded structure is the nucleosome (for reviews, see Kornberg and Klug, 1981; Morse and Simpson, 1988), in which approximately 200 bp of DNA are associated with a histone octamer containing two molecules of each of the four core histones, H2A, H2B, H3 and H4 (Figure 5.5). In

Table 5.1   The histones

Histone	Type	Molecular weight	Molar ratio
H1	Lysine rich	23 000	1
H2A	Slightly lysine rich	13 960	2
H2B	Slightly lysine rich	13 744	2
H3	Arginine rich	15 342	2
H4	Arginine rich	11 282	2

this structure approximately 146 bp of DNA make almost two full turns around the histone octamer and the remainder serves as a linker DNA, joining one nucleosome to another.

The linker DNA is more accessible than the highly protected DNA tightly wrapped around the octamer and is therefore preferentially cleaved when chromatin is digested with small amounts of the DNA-digesting enzyme, micrococcal nuclease. Thus, if DNA is isolated following such mild digestion of chromatin, on gel electrophoresis it produces a ladder of DNA fragments of multiples of 200 bp, representing the results of cleavage in some but not all linker regions (Figure 5.6a, track T). This can be correlated with the properties of nucleosomes isolated from the digested chromatin. Thus a partially digested chromatin preparation can be fractionated into individual nucleosomes associated with 200 bases of DNA (Figure 5.6a, track D), dinucleosomes associated with 400 bases of DNA (Figure 5.6a, track C) and so on (Finch *et al.*, 1975). The individual mononucleosomes, dinucleosomes or larger complexes in each of these fractions can be observed readily in the electron microscope (Figure 5.6b). These experiments provide direct evidence for the organization of DNA into nucleosomes within the cell.

Figure 5.5 Structure of DNA and the core histones within a single nucleosome.

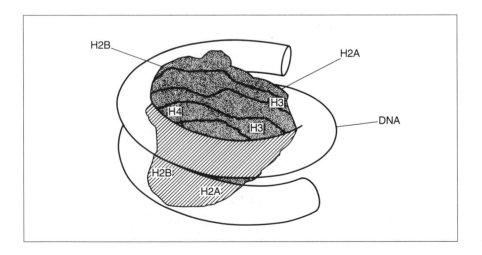

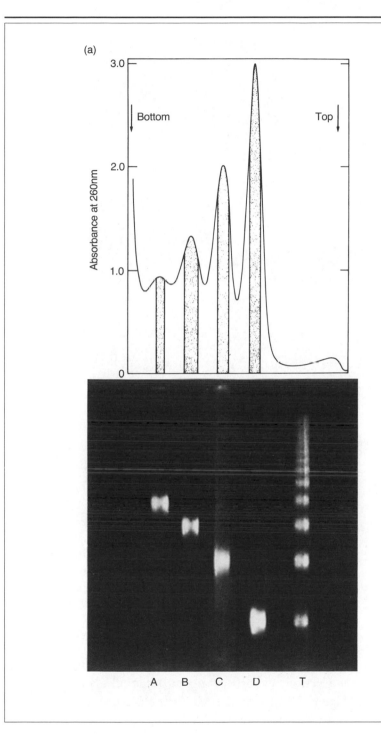

(a)

Figure 5.6 (a) Separation of mononucleosomes (D), dinucleosomes (C), trinucleosomes (B) and tetranucleosomes (A) by sucrose gradient centrifugation. The upper panel shows the peaks of absorbance produced by the individual fractions of the gradient; the lower panel shows the DNA associated with each fraction separated by gel electrophoresis. Track T shows the DNA ladder produced from a preparation containing all the individual nucleosome fractions.

Figure 5.6 (b) Electron microscopic analysis of the fractions separated in panel (a). Mononucleosomes are clearly visible in fraction D, dinucleosomes in fraction C and so on.

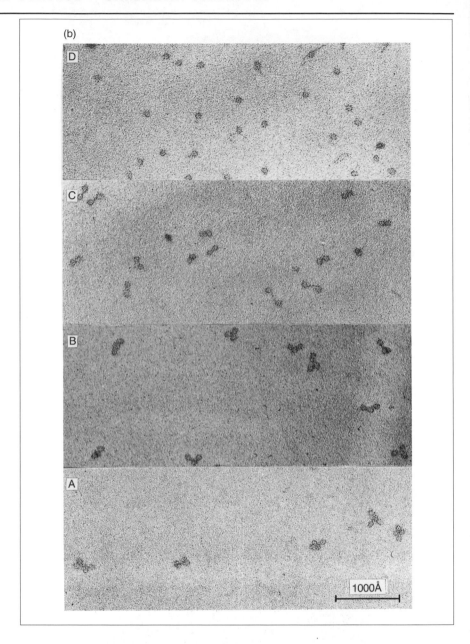

This organization of DNA into nucleosomes, which can be directly visualized in the electron microscope as the beads on a string structure of chromatin (Figure 5.7), constitutes the first stage in the packaging of DNA. Subsequently this structure is folded upon itself into a much more compact structure, known as the solenoid (Figure 5.8). In the formation of this structure histone H1 plays a critical role. As will be seen from Table 5.1, this histone is present at half the level of the other histones and is not

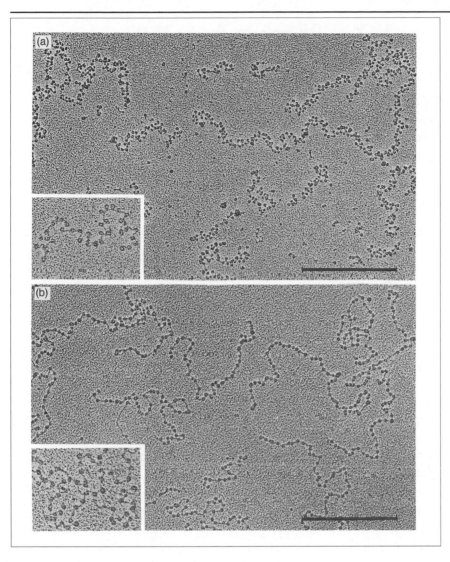

Figure 5.7 Beads on a
string structure of chromatin
visualized in the electron
microscope in the presence
(a) or absence (b) of histone
H1. The bar indicates 0.5 μm.

part of the core histone octamer. Instead, one molecule of histone H1 seals
the two turns which the DNA makes around the core octamer of the
other histones. In the solenoid structure, these single histone H1 molecules associate with one another, resulting in tight packing of the individual nucleosomes into a 30 nm fiber (Thoma *et al.*, 1979; reviewed by
Felsenfeld and McGhee, 1986). This 30 nm fiber is the basic structure of
chromatin in cells not undergoing division, although during cell division
the DNA is further compacted by extensive looping of the 30 nm fiber to
form the readily visible chromosomes. In this structure the fiber loops
are linked at their bases to a protein scaffold known as the nuclear matrix
(Figure 5.9).

Figure 5.8 The solenoid
structure of chromatin.

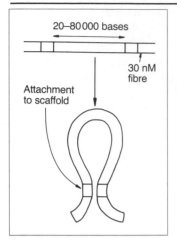

20–80 000 bases

30 nM fibre

Attachment to scaffold

Figure 5.9 Folding of specific regions of the 30 nm solenoid fiber to form a loop which is attached to the nuclear scaffold.

5.4 Changes in chromatin structure in active or potentially active genes

5.4.1 Active DNA is organized in a nucleosomal structure

Having established that the bulk of cellular DNA is associated with histone molecules in a nucleosomal structure, an obvious question is whether genes which are either being transcribed or are about to be transcribed in a particular tissue are also organized in this manner or whether they exist as naked, nucleosome-free DNA.

Two main lines of evidence suggest that such genes are organized into nucleosomes. First, if DNA that is being transcribed is examined in the electron microscope, in most cases the characteristic beads on a string structure (Section 5.3) is observed, with nucleosomes visible both behind and in front of the RNA polymerase molecules transcribing the gene (Figure 5.10; McKnight *et al.*, 1978). Thus, although this structure may break down in genes which are being extremely actively transcribed, such as occurs for the genes encoding ribosomal RNA during oogenesis, it is maintained in most transcribed genes. Secondly, if DNA organized into nucleosomes is isolated as a ladder of characteristically sized fragments following mild digestion with micrococcal nuclease (Figure 5.6), the DNA from active genes is found in these fragments in the same proportion as in total DNA (Lacey and Axel, 1975). Similarly, no enrichment or depletion in the amount of a particular gene found in these fragments is observed when the DNA isolated from a tissue actively transcribing the gene is compared with DNA isolated from a tissue that does not transcribe it. The ovalbumin gene, for example, is found in nucleosome-size fragments of DNA in these experiments regardless of whether chromatin from hormonally stimulated oviduct tissue or from liver tissue is used. This is in agreement with the idea that transcribed DNA is still found in a nucleosomal structure rather than as naked DNA, which would be rapidly digested by micrococcal nuclease and hence would not appear in the ladder of nucleosome-sized DNA fragments.

5.4.2 Sensitivity of active chromatin to DNaseI digestion

In order to probe for other differences that may exist between transcribed and non-transcribed DNA, many workers have studied the sensitivity of these different regions to digestion with the pancreatic enzyme deoxyribonuclease I (DNaseI; reviewed by Weisbrod, 1982; Reeves, 1984). Although this enzyme will eventually digest all the DNA in a cell, if it is applied to chromatin in small amounts for a short period, only a small

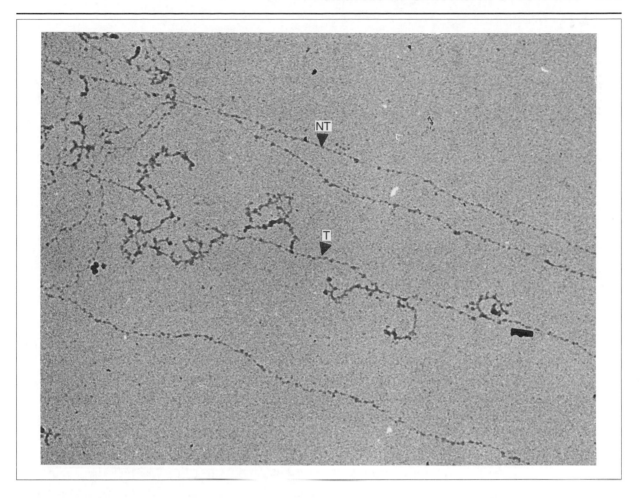

amount of the DNA will be digested. The proportion of transcribed or non-transcribed genes present in the relatively resistant undigested DNA in a given tissue compared with the proportion in total DNA can therefore be used to detect the presence of differences between active and inactive DNA in their sensitivity to digestion with this enzyme. The fate of an individual gene in this procedure can be followed simply by cutting the DNA surviving digestion with an appropriate restriction enzyme and carrying out the standard Southern blotting procedure (Section 2.2.4) with a labeled probe derived from the gene of interest. The presence or absence of the appropriate band derived from the gene in the digested DNA provides a measure of the resistance of the gene to DNaseI digestion (Figure 5.11).

This procedure has been used to demonstrate that a very wide range of genes that are active in a particular tissue exhibit a heightened sensitivity to DNaseI, which extends over the whole of the transcribed gene and for some distance upstream and downstream of the transcribed region. For

Figure 5.10 Electron micrograph of chromatin from a *Drosophila* embryo. Note the identical 'beads on a string' structure of the chromatin that is not being transcribed (NT) and the chromatin that is being transcribed (T) into the readily visible ribonucleoprotein fibrils.

Figure 5.11 Preferential sensitivity of chromatin containing active genes to digestion with DNaseI, as assayed by Southern blotting with specific probes.

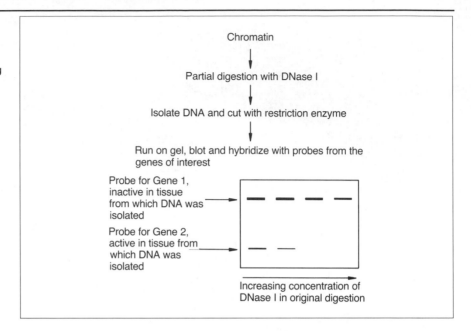

example, when chromatin from chick oviduct tissue is digested, the active ovalbumin gene is rapidly digested and its characteristic band disappears from the Southern blot under digestion conditions which leave the DNA from the inactive globin gene undigested. This difference between the globin and ovalbumin genes is dependent upon their different activity in oviduct tissue rather than any inherent difference in the resistance of the genes themselves to digestion. Thus, if the digestion is carried out with chromatin isolated from red blood cell precursors in which the globin gene is active and that encoding ovalbumin is inactive, the reverse result to that seen in the oviduct is obtained, with the globin DNA exhibiting preferential sensitivity to digestion and the ovalbumin DNA being resistant to such digestion (Stalder *et al.*, 1980a).

It is clear, therefore, that actively transcribed genes, though packaged into nucleosomes, are in a more open chromatin structure than that found for non-transcribed genes and are hence more accessible to digestion with DNaseI. This altered chromatin structure is not confined to the very active genes such as globin and ovalbumin but appears to be a general characteristic of all transcribed genes, whatever their rate of transcription. Thus, if chromatin is digested using conditions that degrade less than 10% of the total DNA, over 90% of transcriptionally active DNA is digested, the DNA of genes encoding rare mRNAs being as sensitive as those encoding abundant mRNAs. Hence, the altered chromatin structure of active genes does not appear to be dependent upon the act of transcription itself, since the genes encoding rare mRNAs will be transcribed only very rarely.

In agreement with this idea, the altered DNaseI sensitivity of previously active genes persists even after transcription has ceased. For example, the ovalbumin gene remains preferentially sensitive to the enzyme when chromatin is isolated from the oviduct after withdrawal of estrogen when, as we have previously seen, transcription of the gene ceases (Section 3.2.3). A similar preferential sensitivity to DNaseI is also observed for the genes encoding the fetal forms of globin, which are not transcribed in adult sheep cells, and for the adult globin genes in mature (14 day) chicken erythrocytes, following the switching off of transcription.

As well as being detectable after transcription has ceased permanently, such increased sensitivity can also be detected in genes about to become active prior to the onset of transcription. As discussed previously (Section 3.2.2), the transcription of the globin gene in Friend erythroleukemia cells only occurs following treatment of the cells with dimethyl sulfoxide. Increased sensitivity to DNaseI digestion is observed, however, in both the treated cells which transcribe the gene at high levels and in the unstimulated cells which do not (Miller *et al.*, 1978).

The altered, more open, chromatin structure detected by increased DNaseI sensitivity does not therefore reflect the act of transcription itself. Rather, it appears to reflect the ability to be transcribed in a particular tissue or cell type. Hence, in cells that have become committed to a particular lineage expressing particular genes, such commitment will be reflected in an altered chromatin structure which will arise prior to the onset of transcription and will persist after transcription has ceased. The ability of imaginal disk cells in *Drosophila* to maintain their commitment to give rise to a particular cell type in the absence of overt differentiation (Section 5.2) is likely, therefore, to be due to the genes required in that cell type having already assumed an open chromatin structure. Moreover, a breakdown in this process resulting in a change in the chromatin structure of a particular regulatory gene would result in a change in the commitment of the disk cells as illustrated in Figure 5.4. Similarly, the altered chromatin structure of the genes required in cartilage cells would be retained in cells cultured in media not supporting expression of these genes, allowing the restoration of the differentiated cartilage phenotype when the cells are transferred to an appropriate medium (Section 5.2).

The alteration of chromatin to a more open structure in committed cells is likely to be a necessary prerequisite for gene expression, allowing the *trans*-acting factors, which actually activate the gene, access to the appropriate sequences within it. Hence the different structure of potentially active genes can explain why a particular steroid hormone, such as estrogen, can induce activity of one particular gene in one tissue and another gene in a different tissue (Section 5.1). Thus the ovalbumin gene in the oviduct would be in an open configuration, allowing induction to occur, whereas the more closed configuration of the vitellogenin gene would not allow access to the complex of hormone and receptor, and induction would not occur. The reverse situation would apply in liver tissue, allowing induction of the vitellogenin and not the ovalbumin gene.

Therefore, the changes in chromatin structure detected by DNaseI digestion play an important role in establishing the commitment to express the specific genes characteristic of a particular lineage, and it is necessary to investigate the mechanisms responsible for this effect.

5.4.3 Mechanism of increased DNaseI sensitivity

If chromatin is digested with low levels of DNaseI, two low molecular weight proteins are released. These proteins are members of the high-mobility group of non-histone proteins associated with DNA, which gain their name from their small size and high proportion of charged amino acids which result in their moving rapidly in gel electrophoresis. These two proteins, known as HMG 14 and 17, play a crucial role in the preferential sensitivity of active DNA to digestion. Thus, if chromatin is treated with 0.35 M sodium chloride, HMG 14 and 17 are released and the preferential sensitivity to digestion of active DNA is lost. The specific pattern of sensitivity characteristic of the particular tissue can be restored, however, by adding back either or both of these proteins. This suggests that HMG 14 and 17 localize in the region of active or potentially active DNA and produce an altered chromatin configuration in this region.

Given that active DNA is organized into nucleosomes (Section 5.4.1), it seems likely that HMG 14 and 17 function by binding to chromatin and preventing the formation of the more tightly folded solenoidal structure (Section 5.3) in which the nucleosomes are packed more tightly, resulting in reduced accessibility of the DNA to DNAseI. Hence, whereas inactive DNA would exist in the solenoidal structure, active or potentially active DNA would be in the more open beads on a string structure.

Histone H1 is known to be essential for the formation of the solenoid structure and the resulting repression of genes within it (for reviews, see Weintraub, 1985; Roth and Allis, 1992) and is known to be depleted in active DNA. Hence HMG 14 and 17 may act by displacing histone H1 from potentially active regions, thereby preventing solenoid formation indirectly. Alternatively, their binding to such regions may simply be

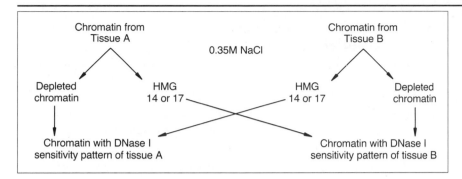

Figure 5.12 Addition of HMG 14 and 17 isolated from one tissue to HMG-depleted chromatin isolated from a second tissue produces the DNaseI sensitivity pattern characteristic of the second tissue.

incompatible with the formation of the solenoid structure, thereby resulting in the more open beads on a string structure.

Whatever the case, the fact that the addition of these proteins to chromatin from which they have been removed restores the preferential sensitivity of active DNA indicates that these proteins are critical for this effect. This finding also forms the basis for distinguishing whether the tissue specificity of this phenomenon, in which genes are only preferentially sensitive in tissues where they will become active, is caused by these proteins or by other features in the chromatin DNA. It is possible to carry out a mixing experiment (Figure 5.12) in which the HMG 14 and 17 proteins isolated from one tissue are mixed with the HMG-depleted chromatin of another tissue. If this is done, the pattern of DNaseI sensitivity produced is that of the tissue from which the depleted chromatin was obtained and not that of the tissue from which the HMG proteins were derived. For example, HMG 14 and 17 isolated from brain tissue can restore the preferential sensitivity of the globin gene when added to depleted red blood cell chromatin, whereas HMG 14 and 17 from red blood cells cannot produce this pattern in brain chromatin (Weisbrod and Weibtraub, 1979).

Hence, although HMG 14 and 17 are critical in producing the open chromatin structure characteristic of active DNA, the tissue-specific differences in the regions of chromatin that are present in this configuration are not paralleled by tissue-specific differences in the nature of HMG 14 and 17. Rather, these proteins are identical in all tissues and must therefore recognize tissue-specific differences in the chromatin DNA, allowing them to bind to particular regions and confer the tissue-specific pattern of DNaseI sensitivity. It is therefore necessary to discuss other features of active chromatin that might be recognized by HMG 14 and 17 or by other regulatory proteins.

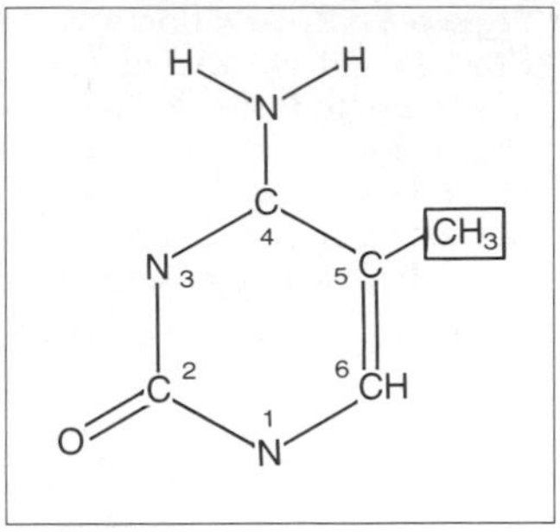

Figure 5.13 Structure of 5-methylcytosine.

5.5   Other changes in DNA and its associated proteins in active or potentially active genes

5.5.1   DNA methylation

Although DNA consists of the four bases adenine, guanine, cytosine and thymine, it has been known for many years that these bases can exist in modified forms bearing additional methyl groups. The most common of these in eukaryotic DNA is 5-methylcytosine (Figure 5.13), between 2 and 7% of the cytosine in mammalian DNA being modified in this way (reviewed by Razin and Cedar, 1991; Bestor and Coxon, 1993).

Approximately 90% of this methylated C occurs in the dinucleotide, CG, where the methylated C is followed on its 3' side by a G residue. Conveniently, this sequence forms part of the recognition sequence (CCGG) for two restriction enzymes, *Msp*I and *Hpa*II, which differ in their ability to cut at this sequence when the central C is methylated. Thus *Msp*I will cut whether or not the C is methylated and *Hpa*II will only do so if the C is unmethylated. This characteristic allows the use of these enzymes to probe the methylation state of the fraction of CG dinucleotides that is within cleavage sites for these enzymes. Hence, if DNA is digested with either *Hpa*II or *Msp*I, both enzymes will give the same pattern of bands only if all the C residues within the recognition sites are unmethylated. In contrast, if any sites are methylated, larger bands will be obtained in the *Hpa*II digest, reflecting the failure of the enzyme to cut at particular sites (Figure 5.14). If this procedure is used in conjunction with Southern blot hybridization using a probe derived from a particular gene, the methyla-

Figure 5.14 Detection of differences in DNA methylation between different tissues, using the restriction enzymes *Msp*I and *Hpa*II.

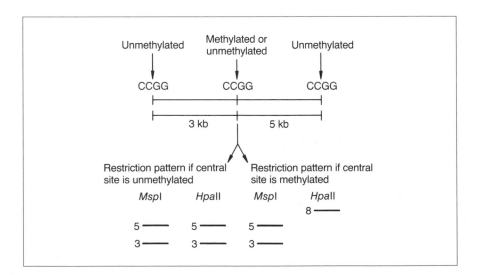

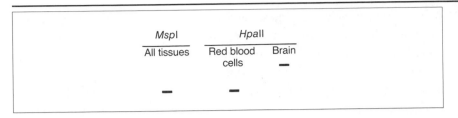

Figure 5.15 Tissue-specific methylation of *Msp*I–*Hpa*II sites in the chicken globin gene results in the methylation-sensitive enzyme, *Hpa*II, producing a band in red blood cell DNA that is identical to that produced by the methylation-insensitive *Msp*I, but producing a larger band in brain DNA.

tion pattern of the *Hpa*II–*Msp*I sites within the gene can be determined.

When this is done it is found that although some CG sites are always unmethylated and others are always methylated, a number of sites exhibit a tissue-specific methylation pattern, being methylated in some tissues but not in others. Such sites within a particular gene are unmethylated in tissues where the gene is active or potentially active and methylated in other tissues (for reviews, see Cedar, 1988; Bird, 1992). For example, a particular site within the chicken globin gene is methylated in a wide variety of tissues and is therefore not digested with *Hpa*I, but is unmethylated and therefore susceptible to digestion in DNA prepared from erythrocytes (Figure 5.15). Similarly, the tyrosine aminotransferase gene, which is expressed only in the liver (Section 5.1), is undermethylated in this tissue when compared with other tissues where it is not expressed. The possibility suggested by this finding that the *trans*-acting factors, which regulate the expression of this gene and are detectable in all tissues, may be inhibited from binding to the gene in tissues where it is methylated, is confirmed by the observation that artificial methylation of the gene prevents the binding of at least one of these factors (Becker *et al.*, 1987).

Interestingly, clusters of unmethylated CG sites known as CpG islands are also observed in constitutively expressed genes, although such sites remain unmethylated in all tissues, consistent with the activity of the gene in all cell types (for review, see Bird, 1987).

In the case of tissue-specific genes which exhibit changes in methylation pattern, under methylation, like DNaseI sensitivity, is observed prior to the onset of transcription and persists after its cessation. For example, the undermethylation of the chicken globin gene persists in mature erythrocytes, where the gene is not being transcribed but is still sensitive to DNaseI digestion. Most interestingly, the region in which unmethylated C residues are found correlates with that exhibiting heightened DNaseI sensitivity (Weintraub *et al.*, 1981) and is also depleted of histone H1 (Ball *et al.*, 1983). As with DNaseI sensitivity, therefore, undermethylation is a consequence of commitment to a particular pattern of gene expression and is associated with the change in chromatin structure observed in active or potentially active genes.

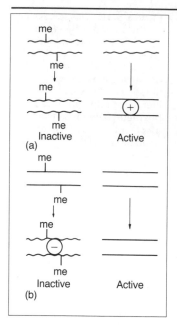

Figure 5.16 The transition from an inactive (wavy line) to an active state (straight line) of the DNA may take place via an activating protein which binds specifically to unmethylated DNA, thereby activating it (a), or via an inhibitory protein which binds specifically to methylated DNA, thereby repressing it (b).

It is clearly possible that these methylation differences between inactive or potentially active genes can be recognized by proteins, since the proteinaceous enzyme *Hpa*II digests only unmethylated DNA (see above). Similarly the effect of a T to C mutation in the *lac* operon, which prevents binding of the *lac* repressor protein, can be suppressed by methylating the C residue to yield 5-methylcytosine. Differences in methylation levels in the same gene in different tissues could therefore result in differences in recognition by specific proteins. This in turn could lead to the changes in chromatin structure which result in the observed changes in DNaseI sensitivity. Thus it is possible that undermethylation promotes the binding of proteins such as HMG14 and 17 which produce a more open chromatin structure (Figure 5.16a). Alternatively, an inhibitory protein may bind preferentially to methylated DNA, resulting in gene repression (Figure 5.16b). This latter possibility is supported by the identification of a specific protein (MeCP-1) which binds preferentially to methylated DNA and which can repress gene expression (Boyes and Bird, 1991).

The idea that methylation differences might be the essential feature causing altered chromatin structure is particularly attractive because of the ease with which such differences can be propagated, allowing cellular commitment to be stable over many generations (Section 5.1). Thus, in double-stranded DNA, the CG dinucleotide will exist as a symmetrical structure:

$$5'\text{-CG-}3'$$
$$3'\text{-GC-}5'$$

It has been observed that when one C in this structure is methylated, the C on the opposite strand is also methylated. This effect is achieved by a DNA methyltransferase enzyme (for review, see Kumar *et al.*, 1994). This enzyme recognizes sites where only one C is methylated (hemimethylated sites) and methylates the second C residue rapidly. Hence, the pattern of DNA methylation will be maintained following DNA replication, with the hemimethylated sites produced by replication being re-methylated rapidly (Figure 5.17). Similarly, because the maintenance methylase is only active on hemimethylated sites, unmethylated sites present in a particular tissue will be propagated through subsequent cell divisions. Thus, once established, a particular pattern of methylation will be maintained, accounting for the stability of the committed state.

Such a mechanism also allows readily for the specific loss of methylation sites that must occur during the process of commitment to a particular lineage. Thus, such losses could occur via a specific demethylation event or simply by inhibiting the action of the maintenance methylase at a partic-

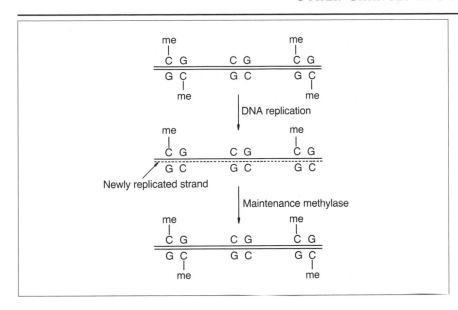

Figure 5.17 Model for the propagation of methylation patterns through cell division.

ular site following cell division (Figure 5.18). This latter mechanism would eventually result in the generation of one daughter cell in which the site was fully methylated and one in which it was unmethylated. As we have seen (Sections 2.2.2 and 2.4), such a differentiation event in which a stem cell divides to yield one daughter that differentiates and another that maintains the stem cell lineage is a common feature of embryonic development. As with other methylation patterns, the new demethylated site in the committed cell would be propagated through subsequent cell divisions.

Hence DNA methylation processes provide a means of explaining the stability of the committed state, while allowing for its modification in suitable circumstances. Unlike DNA deletion events, methylation patterns are not irreversible and could be altered when a cell undergoes transdifferentiation or following nuclear transplantation (Section 2.2.3). As we have discussed, however, such changes in the differentiated state normally require dedifferentiation and cell division, exactly as would be necessary for a process dependent on DNA replication and subsequent inhibition of maintenance methylation at particular sites.

A model in which the regulation of methylation at particular sites controls cellular commitment and the opening of chromatin around active genes is therefore an attractive one. However, the correlation of under-methylation with the ability of a gene to be transcribed is insufficient to prove the truth of this model, and it is necessary to consider other more direct evidence in its favor. This evidence comes from two areas of investigation.

Figure 5.18 Alteration of the methylation pattern to produce an unmethylated site by inhibition of the maintenance methylase and subsequent DNA replication.

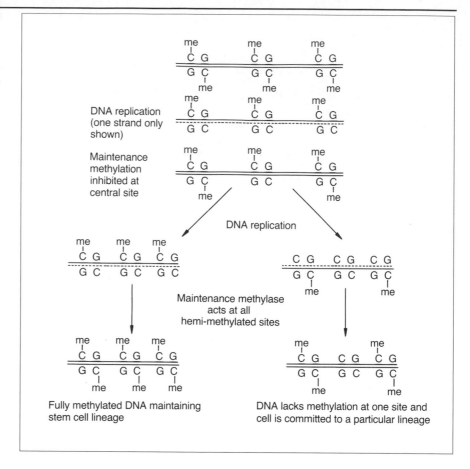

Introduction of methylated DNA into cells

A number of experiments (reviewed by Bird, 1984) have shown that if DNA containing 5-methylcytosine is introduced into cells it is not expressed, whereas the same DNA which has been demethylated is expressed. These experiments have been carried out using both eukaryotic viruses and cellular genes, such as those encoding β- and γ-globin, cloned into plasmid vectors. In these experiments the methylated DNA adopts a DNaseI-insensitive structure typical of inactive genes whereas unmethylated DNA adopts the DNaseI-sensitive structure typical of active genes (Figure 5.19; Keshet *et al.*, 1986), providing direct evidence for the role of methylation differences in regulating the generation of different forms of chromatin structure.

Effect of artificially induced demethylation

If methylation differences play a crucial role in the regulation of differentiation, it should be possible to change gene expression by demethylating

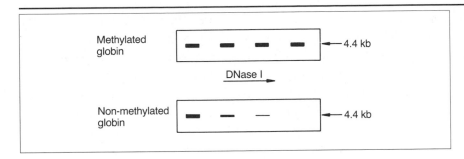

Figure 5.19 Unmethylated DNA introduced into cells adopts an open DNaseI-sensitive configuration, whereas the same DNA when methylated and then introduced into cells adopts a more tightly packed DNaseI-insensitive form.

DNA. This has been achieved in a number of cases by treating cells with the cytidine analog, 5-azacytidine, which is incorporated into DNA but cannot be methylated, having a nitrogen atom instead of a carbon atom at position 5 of the pyrimidine ring (for review, see Cedar, 1988). In the most dramatic of these cases, treatment of an undifferentiated fibroblast cell line with this compound results in the activation of a few key regulatory loci and the cells differentiate into multinucleate, twitching, striated muscle cells (Constantinides *et al.*, 1977) (see also Section 7.2.4). In other cases, although not actually producing altered gene expression, demethylation may facilitate it. Thus, if HeLa cells are treated with 5-azacytidine no dramatic effects are observed. If such cells are fused with muscle cells, however, muscle-specific genes are switched on in the treated HeLa cells, a phenomenon which is not observed when untreated HeLa cells are fused with mouse muscle cells (Chiu and Blau, 1985). Hence, treatment of the HeLa cells has altered their muscle-specific genes in such a way as to allow them to respond to *trans*-acting factors present in the mouse muscle cells. This type of regulatory process is exactly what would be predicted from the role of methylation in the alteration of chromatin structure and thereby in facilitating interactions with *trans*-acting regulatory factors. It should be noted, however, that in none of these cases has it been shown directly that 5-azacytidine achieves its effect by inducing demethylation rather than by some other, as yet uncharacterized, action of this compound.

Despite this caveat, the available evidence from these studies, and those discussed above, indicates that DNA methylation plays a central role in the regulation of gene expression, at least in mammals. This conclusion is reinforced by the finding that inactivation of the gene encoding the DNA methyltransferase which methylates DNA results in the death of mouse embryos prior to birth (Li *et al.*, 1992). Hence at least in mammals DNA methylation is essential for normal embryonic development.

It is noteworthy, however, that *Drosophila* does not contain 5-methyl-cytosine (Uriel-Shoval *et al.*, 1982) and no clear example of a methylated gene has been detected in an invertebrate, although DNaseI sensitivity changes occur in these organisms exactly as in vertebrates. Hence, if methylation of cytosine is indeed the primary cause of changes in chromatin structure in vertebrates, other mechanisms must produce the same effect in invertebrates. Such mechanisms may also be involved in the minority of cases in vertebrates, such as the chicken α 2 (I) collagen gene, where differences in methylation between expressing and non-expressing tissues cannot be detected (McKeon *et al.*, 1982). Other features of active chromatin which distinguish it from inactive chromatin and which might be involved in this process will now be discussed.

5.5.2 Histone modifications

Given the essential role of histones in chromatin structure, it is possible that potentially active chromatin might be marked in some way by modification of the histones within it. In fact a number of such modifications of these proteins (involving, for example, methylation, phosphorylation, ubiquitination, or acetylation) have been reported (for review, see Turner, 1993). However, we shall discuss only two of these, which have been correlated with active or potentially active regions of chromatin.

Ubiquination

Ubiquitin is a small protein of only 76 amino acids, which forms a conjugate with histone H2A in which the C-terminal carboxyl group of ubiquitin is joined to the free amino group on an internal lysine residue in the histone, to form a branched molecule (Figure 5.20; reviewed by Busch and Goldknopf, 1981). This modification reduces the net positive charge on the histone molecule, both by neutralizing the charged amino group on lysine and by introducing a number of negatively charged amino acids present in the ubiquitin molecule itself. Only a small minority of the H2A in a cell (about 5–10%) exists in this form but, in *Drosophila* at least, this modified form of H2A is localized preferentially in

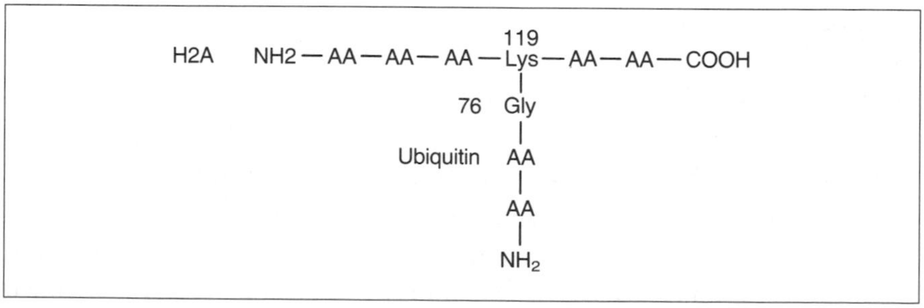

Figure 5.20 Linkage of ubiquitin to histone H2A. The carboxyl terminal amino acid (76) of ubiquitin links to the lysine at position 119 of histone H2A. AA indicates the amino acid backbone of the molecules.

```
                              119
H2A    NH2 — AA — AA — AA — Lys — AA — AA — COOH
                               |
                        76   Gly
                               |
         Ubiquitin      AA
                               |
                        AA
                               |
                        NH2
```

nucleosomes containing active genes (Levinger and Varshavsky, 1982). Hence, it is possible that ubiquitination may substitute for DNA methylation in regulating chromatin structure in invertebrates and also in some cases in vertebrates. Unlike DNA methylation, however, the putative role of ubiquitination is based only on the correlation described above, no direct evidence that H2A ubiquitination actually affects gene expression being available.

Acetylation

The free amino group on internal lysine residues is also involved in the second modification of histones found in active or potentially active DNA, i.e. acetylation (for review, see Turner, 1993). In this case, however, one of the hydrogen atoms in the free amino group is substituted by an acetyl group (CH_3CO). This modification, which like ubiquitination reduces the net positive charge on the histone molecule, occurs primarily for histones H3 and H4. Hyperacetylated forms of these histones, containing several such acetyl groups, have been shown to be localized preferentially in active genes exhibiting DNaseI sensitivity whilst under-acetylation of the histones is characteristic of transcriptionally inactive regions. Furthermore, treatment of cells with sodium butyrate, which inhibits a cellular deacetylase activity and hence increases histone acetylation, has been shown to result in DNaseI sensitivity of some regions of chromatin and to activate the expression of some previously silent cellular genes (Reeves and Cserjesi, 1979). Hence, as with DNA methylation, there is direct evidence that hyperacetylation of histones may play some role in the generation of DNaseI sensitivity and the marking of active or potentially active genes.

The effects of histone modifications such as ubiquitination or acetylation on chromatin structure may operate in one of two possible ways (Figure 5.21). Firstly, such changes might act by altering the charge distribution on the histone molecules, thereby affecting their association with each other or with the DNA. In turn this would result in improved access to the DNA for factors which can stimulate transcription (Figure 5.21a). Some evidence for this model has been provided in the case of histone acetylation where the binding to DNA of the *Xenopus* transcription factor TFIIIA was greatly enhanced when the nucleosomes contained hyperacetylated histone H3 and H4 compared with when these proteins were not acetylated.

An alternative possibility is that these modifications affect the protein–protein interaction of the histones with other regulatory molecules. This would parallel the role proposed for methylation differences in

Figure 5.21 Acetylation (A) of histones (H) may activate gene expression either by directly altering the conformation of the histones themselves, producing a more open chromatin structure (a), or indirectly by disrupting their association with an inhibitory molecule (X) (b).

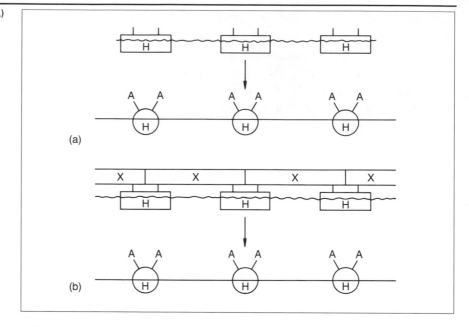

(a)

(b)

affecting the binding of positively or negatively acting factors to the DNA (Figure 5.16; Section 5.5.1). Thus, for example, ubiquitinated or acetylated histones might be recognized by HMG14 and 17, thereby acting as markers guiding HMG14 and 17 to certain regions and therefore leading indirectly to the destabilization of the solenoid structure. Similarly, ubiquitination or acetylation could disrupt an association with an inhibitory molecule involved in maintaining the closed chromatin structure (Figure 5.21b). An example of this is seen in yeast where mutation of the lysine at position 16 in histone H4 to a neutral amino acid such as glycine results in constitutive expression of the normally silent mating type loci (Section 2.4), leading to drastically reduced mating efficiency. This mutation can be suppressed by a compensating mutation in the SIN3 gene, whose protein product is normally required for repression of the mating type genes. This suggests that SIN3 may normally interact with histone H4 via the positively charged lysine residue and repress the mating type loci. Normally acetylation of the lysine removes the positive charge, hence leading to dissociation of SIN3 and allowing controlled activation of the mating type loci. However, substitution of a neutral amino acid prevents the interaction of H4 and SIN3 from occurring, thereby resulting in constitutive activation, unless a compensating mutation is introduced into SIN3 allowing it to bind to H4 in the absence of a positively charged residue. It is possible, therefore, that histone modifications can affect chromatin structure either directly or indirectly via interactions with other proteins. Regardless of the precise mechanism by which they act, however, it is likely that such modifications do play an important role in the changes in chromatin structure which are necessary for gene activation to occur.

5.6 DNaseI-hypersensitive sites in active or potentially active genes

5.6.1 Detection of DNaseI-hypersensitive sites

So far in this chapter we have seen that the region of chromatin containing an active or potentially active gene has a number of distinguishing features, including undermethylation, histone modifications and increased sensitivity to digestion with DNaseI. Such changes extend over the entire region of the gene and some flanking sequences and, in the case of DNaseI sensitivity, result in an approximately 10-fold increase in the rate at which active or potentially active genes are digested.

Following the discovery of such increased DNaseI sensitivity, many investigators studied whether within the region of increased sensitivity there might be particular sites which were even more sensitive to cutting with the enzyme and which would therefore be cut even before the bulk of active DNA was digested. The technique used to look for such sites is based on that used to look at the overall DNaseI sensitivity of a particular region of DNA (Section 5.4.2). Chromatin is digested with DNaseI and a restriction enzyme and then subjected to a Southern blotting procedure using a probe derived from the gene of interest. As we have seen previously, the overall sensitivity of the gene can be monitored by observing how rapidly the specific restriction enzyme fragment derived from the gene disappears with increasing amounts of the enzyme (Figure 5.11). To search for hypersensitive sites, however, much lower concentrations of the enzyme are used and the appearance of discrete digested fragments derived from the gene is monitored (Figure 5.22). Such specific fragments have at one end the cutting site for the restriction enzyme used and, at the other, a site at which DNaseI has cut, producing a defined fragment. Since the position at which the restriction enzyme cuts in the gene is known, the position of the hypersensitive site can be mapped simply by determining the size of the fragment produced.

Using this procedure a very wide variety of genes have been shown to contain such hypersensitive sites exhibiting a sensitivity to DNaseI digestion 10-fold above that of the remainder of an active gene and therefore about 100-fold above that seen in inactive DNA (for review, see Gross and Garrard, 1988). A representative list of cases in which such sites have been

Figure 5.22 Detection of DNaseI-hypersensitive sites in active genes by mild digestion of chromatin to produce a digestion product with a restriction site at one end and a DNaseI-hypersensitive site at the other (right-hand panel). More extensive digestion will result in the disappearance of the band (central panel) as in the experiment illustrated in Figure 5.11.

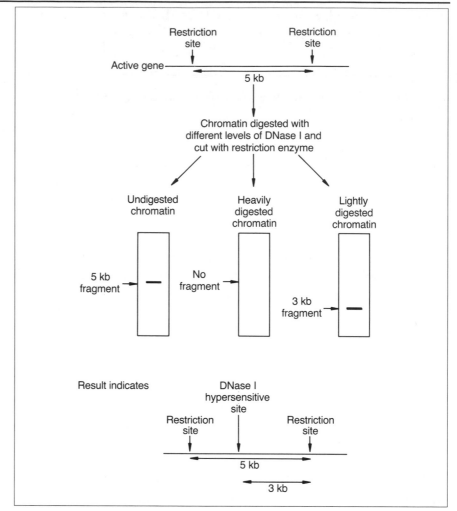

detected is given in Table 5.2. As with the increased sensitivity of the gene itself, many hypersensitive sites appear only in tissues where the gene is active. Thus, the increased sensitivity of globin DNA in erythrocytes to digestion is paralleled by the presence of hypersensitive sites within the gene in erythrocyte chromatin but not in that of other tissues (Stalder *et al.*, 1980b). Similarly, the ovalbumin gene in hormonally treated chick oviduct also exhibits hypersensitive sites that are not found in other tissues, including erythrocytes (Kaye *et al.*, 1984; Figure 5.23).

As with undermethylation and the sensitivity of the entire gene to digestion, DNaseI-hypersensitive sites appear to be related to the potential for gene expression rather than always being associated with the act of transcription itself. Thus the hypersensitive sites near the *Drosophila* heat-shock genes are present in the chromatin of embryonic cells prior to any

Table 5.2   Examples of genes containing DNaseI-hypersensitive sites

(a) *Tissue-specific genes*

Immune system	Immunoglobulin, complement C4
Red blood cells	α-, β- and ε-globin
Liver	α-fetoprotein, serum albumin
Nervous system	Acetylcholine receptor
Pancreas	Preproinsulin, elastase
Connective tissue	Collagen
Pituitary gland	Prolactin
Salivary gland	*Drosophila* glue proteins
Silk gland	Silk moth fibroin

(b) *Inducible genes*

Steroid hormones	Ovalbumin, vitellogenin, tyrosine aminotransferase
Stress	Heat-shock proteins
Viral infection	β-interferon
Amino acid starvation	Yeast HIS 3 gene
Carbon source	Yeast GAL genes, yeast ADH 1 gene

(c) *Others*

Histones, ribosomal RNA, 5S RNA, transfer RNA, cellular oncogenes c-*myc* and c-*ras*, glucose-6-phosphate dehydrogenase, dihydrofolate reductase, cysteine protease, etc.

heat-induced transcription of these genes (Keene *et al.*, 1981), and one of the sites in the mouse α-fetoprotein gene persists in the chromatin of adult liver after the transcription of the gene (which is confined to the fetal liver) has ceased (Nahon *et al.*, 1987).

Hence, as with DNA methylation and generally increased sensitivity to DNaseI, the appearance of hypersensitive sites appears to be involved in gene regulation. This idea is reinforced by the location of the hypersensitive sites, which can be precisely mapped as described above. Many sites

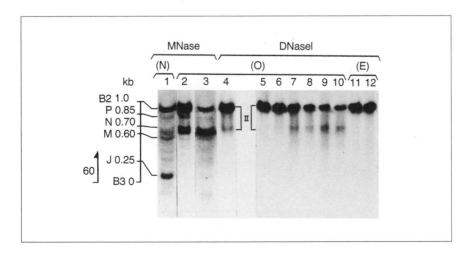

Figure 5.23 Detection of a DNaseI-hypersensitive site in the ovalbumin gene in oviduct tissue (O) but not in erythrocytes (E). Track 4 shows the detection of a lower band caused by cleavage at a hypersensitive site when oviduct chromatin is digested with DNaseI. *Note the progressive appearance of this band as increasing amounts of DNaseI are used to digest the oviduct chromatin (tracks 5–10). No cleavage is observed when similar amounts of DNaseI are used to cut erythrocyte chromatin (tracks 11 and 12). The hypersensitive site in oviduct chromatin is also cleaved, however, with micrococcal nuclease (tracks 2 and 3). Track 1 shows the pattern produced by micrococcal nuclease cleavage of naked DNA (N).

are located at the 5' end of the genes, in positions corresponding to DNA sequences that are known to be important in regulating transcription. For example, a site present at the 5' end of the steroid-inducible tyrosine aminotransferase gene is localized within the DNA sequence that is responsible for the steroid inducibility of the gene (Section 6.2.3). Even in cases where hypersensitive sites are located far from the site of transcriptional initiation, they appear to correspond to other regulatory sequences, such as enhancers, which can act over large distances (Section 6.3).

In the case of the *Drosophila* gene encoding the glue protein Sgs4, the fortuitous existence of a mutant strain of fly has indicated the functional importance of hypersensitive sites. In normal flies this gene contains two hypersensitive sites, 405 and 480 bases upstream of the start of transcription. In the mutant, both sites are removed by a small DNA deletion of 100 bp. Despite the fact that this gene still has the start site of transcription and 350 bases of upstream sequences, no transcription occurs (Figure 5.24), indicating the regulatory importance of the region containing the hypersensitive sites (Shermoen and Beckendorf, 1982).

It is clear, therefore, that hypersensitive sites represent another marker for active or potentially active chromatin and are a feature likely to be of particular importance in gene regulation, being associated with many DNA sequences that regulate gene expression. It is therefore necessary to consider the nature and significance of these sites.

5.6.2   Nature and significance of hypersensitive sites

In many cases DNaseI-hypersensitive sites are also exquisitely sensitive to cleavage with other enzymes, such as the S1, *Bal*31 and micrococcal nucleases (Figure 5.23; reviewed by Elgin, 1984), which are particularly active on DNA that is single stranded or has an altered configuration. Although it is unlikely that the DNA in hypersensitive sites is actually single stranded, since it is also cut by enzymes specific for double-stranded

Figure 5.24 Deletion of a region containing the two hypersensitive sites upstream of the *Drosophila sgs4* gene abolishes transcription.

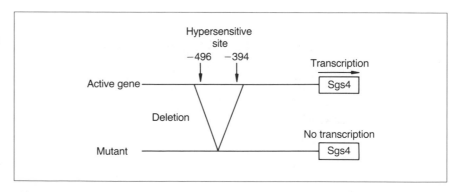

DNA, it is likely that it exists in a configuration different from that of the bulk DNA. This may involve, for example, the DNA in this region being in a highly supercoiled form which is under torsional stress. Such super-coiling has been shown to stabilize an alternative structural form of DNA, known as Z-DNA, in which the DNA helix coils in a left-handed rather than the conventional right-handed manner (for review, see Rich *et al.*, 1984).

A variety of evidence suggests that the formation of Z-DNA is associat-ed with transcriptional activity. Thus antibodies that specifically recognize Z-DNA are known to stain the heat-inducible puffs in *Drosophila* poly-tene chromosomes (Pardue *et al.*, 1983), as well as the transcriptionally active macronucleus of the ciliated protozoan *Stylonichia*. Similarly, sites that form Z-DNA (consisting of alternating purines and pyrimidines) preferentially are associated closely with DNaseI-hypersensitive sites in a number of different situations, e.g. within the enhancer element that regulates transcription of the eukaryotic virus SV40 (Figure 5.25).

When this virus enters cells its DNA, which is circular and only 5000 bases in size, becomes associated with histones in a typical nucleosomal structure, which can be visualized in the electron microscope as a mini-chromosome (Saragosti *et al.*, 1980; Figure 5.26). When this is done, however, the region containing the hypersensitive sites and potential Z-DNA-forming regions remains nucleosome free and is seen as naked DNA. A similar lack of nucleosomes in the region of hypersensitive sites is also found in the chicken β-globin gene, the 5' hypersensitive site of this gene being excisable as a 115 bp restriction fragment lacking any associat-ed nucleosomes (McGhee *et al.*, 1981).

It is clear, therefore, that hypersensitive sites represent regions of DNA which are free of nucleosomes and which may have an unusual configura-tion. Such a non-nucleosomal structure is likely to facilitate the entry of regulatory proteins or RNA polymerase itself into the DNA and therefore to allow the onset of transcription. In agreement with this idea, studies on the interaction of RNA polymerase with chromatin in the test tube have shown that the enzyme cannot initiate transcription on DNA that is

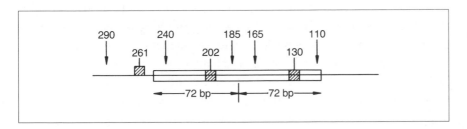

Figure 5.25 Association of hypersensitive sites (arrows) with regions of Z-DNA (shaded) in the DNA of the SV40 enhancer element. The boxed regions indicate the 72 bp sequence, which is repeated twice in the enhancer. The numbers indicate the position of each element within the enhancer.

Figure 5.26 Electron micrograph of the SV40 mini-chromosome consisting of DNA and associated histones. Note the region of the enhancer and hypersensitive sites, which appears as a thin filament of DNA free of associated proteins.

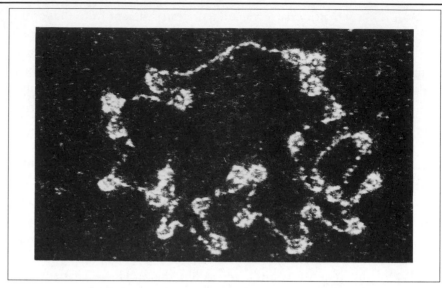

packaged in nucleosomes, although as previously discussed (Section 5.4.1) it can elongate a previously initiated transcript through a region that is packaged in this way (Lorch *et al.*, 1987).

Hence, the existence of a nucleosome-free site is probably essential for the onset of transcription (for review, see Felsenfeld, 1992). As with the general sensitivity of active genes to DNaseI or undermethylation, the existence of such a site is likely to be necessary but not sufficient for transcription, which will require the binding to the DNA of other *trans*-acting factors or of RNA polymerase itself. In the case of the heat-shock genes whose transcription is stimulated by elevated temperatures (Section 3.2.4) hypersensitive sites are present prior to heat treatment. Following heat treatment, a protein factor, known as the heat-shock factor (HSF), binds to this region of DNA and transcription is stimulated (Section 6.2.2). In this case, the HSF is only capable of binding following heat shock and hence stimulation of transcription only occurs following such treatment (Figure 5.27a). In other cases, however, where the necessary transcription factors are present in all tissues, transcription may follow immediately the nucleosome-free region is generated, allowing these factors access. Thus, in the case of glucocorticoid-responsive genes (reviewed by Beato, 1989), the critical regulatory event is the binding of the glucocorticoid receptor–steroid complex to a particular DNA sequence, which displaces a nucleosome and generates a DNaseI-hypersensitive site. Ubiquitous factors present in all tissues, such as NFI and the TATA box binding factor, immediately bind to this region and transcription begins (Figure 5.27b).

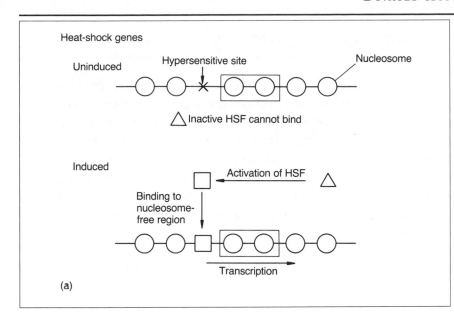

(a)

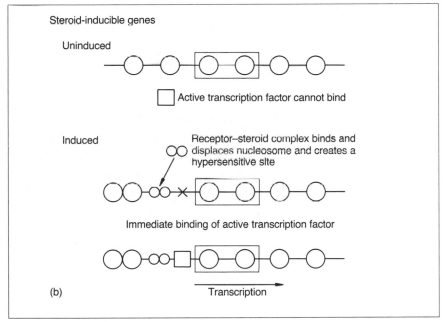

(b)

Figure 5.27 Two mechanisms for transcriptional activation. (a) The heat-shock transcription factor (HSF) is activated by heat and binds to a pre-existing nucleosome-free region. (b) The receptor–steroid complex displaces a nucleosome, creating a hypersensitive site, and allowing an active transcription factor to bind. X indicates the position of a hypersensitive site.

Although these two situations appear different in terms of the time at which the hypersensitive site appears relative to the onset of transcription, they illustrate the basic role of hypersensitive sites, i.e. the displacement of nucleosomes and the generation of a site of access for regulatory proteins (for review, see Workman and Buchman, 1993).

5.7   Situations in which chromatin structure is regulated

In this chapter we have discussed the role of changes in the chromatin structure of individual genes in mediating commitment to a particular differentiated state and thereby allowing the tissue-specific activation of gene expression. However, differences in chromatin structure are also involved in regulating gene expression in two other well-characterized processes, i.e. X chromosome inactivation and genomic imprinting. Both of these processes involve differences in expression between the two copies of a specific gene which are present on different homologous chromosomes in a diploid cell (for review, see Lyon, 1993).

5.7.1   X chromosome inactivation

The fact that females of mammalian species have two X chromosomes whereas males have one X and one Y chromosome creates a problem of how to compensate for the difference in dosage of genes on the X chromosome which occurs because females have two copies whereas males have only one. In mammals this problem is solved by the process of X chromosome inactivation (for reviews, see Grant and Chapman, 1988; Lyon, 1992).

Thus during the process of embryonic development one of the two X chromosomes in each female cell undergoes an inactivation process so that the expression of virtually all the genes on this chromosome is inactivated whilst those on the other chromosome remain active. This results in each female cell having only one active copy of X chromosome genes, paralleling the situation in male cells which have only one X chromosome (Figure 5.28).

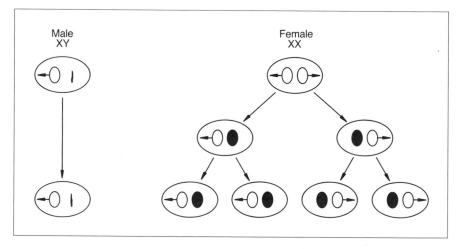

Figure 5.28 X chromosome inactivation results in one or other of the two X chromosomes becoming inactivated (solid) in each cell whilst the other remains active (open).

This process occurs apparently randomly, with cells in the early female embryo inactivating either the X chromosome inherited from the father (the paternal chromosome) or that inherited from the mother (the maternal chromosome). However, once one or other of the X chromosomes has been inactivated, this inactivation is propagated stably through cell division to all the progeny of that cell (Figure 5.28). Such stability is dependent upon the fact that the inactive and active X chromosomes have a different chromatin structure.

Thus within the inactive X chromosome the DNA is tightly packed into a highly condensed structure known as heterochromatin and this high-density structure results in the inactive X chromosome being visible as a distinct element within the cell known as a Barr body. In turn such a condensed structure has been shown to involve decreased sensitivity to digestion with DNaseI, reduced histone acetylation and enhanced methylation which can be observed when genes on the inactive X chromosome are compared with the equivalent gene on the active chromosome. Thus, for example, 60 of the 61 CpG dinucleotides in the CpG island (Section 5.5.1) located around the promoter of the PGKI gene are methylated on the C residue when the inactive X chromosome is studied, whereas all these sites are unmethylated on the active X. Moreover, treatment with 5-azacytidine, leading to demethylation (Section 5.5.1), can reactivate previously inactivated X chromosome genes. Interestingly, in *Drosophila* compensation for the reduced number of X chromosomes in male cells is achieved by increasing the transcriptional activity of the genes on the single male X chromosome rather than by inactivating one of the female X chromosomes. As in the case of X inactivation, however, this effect appears to involve alterations in chromatin structure since the male X chromosome has a higher level of acetylated histones than either of the female chromosomes (Kuroda *et al.*, 1993).

Chromatin structure thus plays a critical role in differentially regulating the activity of genes on the X chromosomes in females and males and in particular in the maintenance of X chromosome inactivation through cell division. Interestingly the onset of X chromosome inactivation in the embryo requires a particular region of the chromosome known as the X inactivation center. If this region is deleted then X inactivation does not occur. A gene known as XIST has been mapped to the X inactivation center. Paradoxically, the XIST gene is transcribed only on the inactive X chromosome and not on the active chromosome, the opposite pattern to all other genes (for review, see Ballabio and Willard, 1992). It is possible therefore that transcription of XIST on one of the two X chromosomes early in development in some way results in a change in chromatin

Figure 5.29 Transcription of the XIST gene on one of the two X chromosomes is associated with its inactivation via a change in chromatin structure.

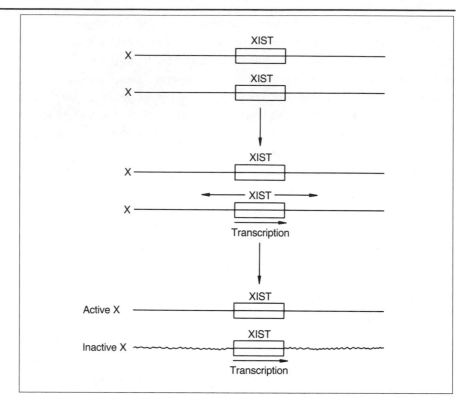

structure which is propagated along the rest of the chromosome, switching off all other genes (Figure 5.29).

Although this mechanism for the onset of X chromosome inactivation remains speculative it is clear that the initiation and maintenance of this process requires the ability to produce and stably propagate an altered chromatin structure which we have previously seen to be critical in tissue-specific gene regulation.

5.7.2   Genomic imprinting

Genomic imprinting (for reviews, see Lyon, 1993; Surani, 1991) resembles X chromosome inactivation in that one of the two copies of specific genes is inactivated whilst the other remains active. This process differs, however, in that only a few genes scattered on different autosomes (non-sex chromosomes) are inactivated and that the same copy is always inactivated in all cells and in all organisms whether male or female. Thus of the first five genes which have been shown to be imprinted, it is always the maternally inherited copy of the genes encoding the insulin-like growth factor 2 (IGF2) protein, the SmN splicing protein (Section 4.2.3) and the U2AF35-related protein which are inactivated, with the paternally inherited gene remaining active. Conversely the paternally inherited copies of

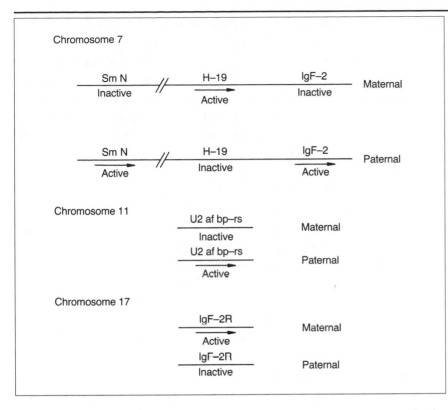

Figure 5.30 Imprinting results in the inactivation of the maternally inherited SmN, U2AFbp-rs and Igf-2 genes and of the paternally inherited Igf-2R and H19 genes.

the IGF2 receptor gene and the H19 gene are inactivated, with the maternally inherited gene remaining active (Figure 5.30) (Barlow, 1993; Hayashizaki *et al.*, 1994).

As with X inactivation, therefore, this process results in all cells having only one functional copy of each of these genes, although in the case of genomic imprinting all cells express the same copy. This process cannot be reversed during embryonic development even when genetic crosses are used to produce embryos with two imprinted copies of the gene. Thus embryos which inherit two maternal and no paternal copies of the IGF2 gene or two maternal but no paternal copies of the SmN gene die due to the lack of an active gene. These embryos therefore die due to the absence of a functional protein even though two copies of the gene capable of encoding this protein are present in each cell of the embryo.

Despite the lethal effects which can result when imprinting goes wrong, it is unclear what the normal function of this process is. It has been suggested, for example, that imprinting may represent a means of preventing the parthenogenetic development of the unfertilized egg to produce a haploid embryo with no paternal contribution. Alternatively, it may have evolved because of the conflict between the maternal and paternal genomes in terms of the transfer of nutrients from mother to offspring.

Thus whilst it is in the paternal interest to promote the growth of the individual fetus the maternal interest is to restrict the growth of any individual fetus so that other fetuses fathered by different males either concurrently (in multi-fetal litters) or subsequently can develop fully (for discussion, see Moore and Haig, 1991). This idea is in agreement with the involvement of at least two imprinted genes (IGF2 and its receptor) in growth regulatory processes but there is no direct evidence to support it. Interestingly, however, the proteins encoded by two other imprinted genes, SmN and the U2AF35-related protein, encode factors which appear to be involved in RNA splicing (Section 4.2) and have no obvious connection with growth regulation.

This lack of a clear functional role for genomic imprinting is in contrast to the clear role of X chromosome inactivation in compensating for the extra X chromosome in females compared with males. Indeed it has even been suggested that imprinting may not have a function at all but may represent a vestige of an evolutionarily ancient defense system to inactivate foreign DNA, with the imprinted genes having some feature which causes this system to consider them as foreign (see Barlow, 1993, for further discussion of this idea).

Whatever its precise function (if any), it is clear that genomic imprinting resembles X chromosome inactivation in that specific differences exist in the methylation pattern between the active and inactive copies of the imprinted gene. Moreover, it has recently been shown that embryos which lack a functional DNA methyltransferase and which therefore cannot methylate their DNA (Section 5.5.1) fail to carry out genomic imprinting (for reviews, see Surani, 1993; Reik and Allen, 1994). Hence methylation of one of the two copies of the gene is critical for imprinting to occur.

5.8 Conclusions

A variety of changes take place in the chromatin of genes during the process of commitment to a particular pathway of differentiation. Such changes involve both modification of the DNA itself by undermethylation, to the histones with which it is associated and to the general packaging of the DNA in chromatin. They result in three levels of chromatin structure within the cell (Figure 5.31). Thus although the bulk of inactive DNA is organized into a tightly packed solenoid structure, active or potentially active genes are organized into a more open 'beads on a string' structure and short regions within the gene are present in

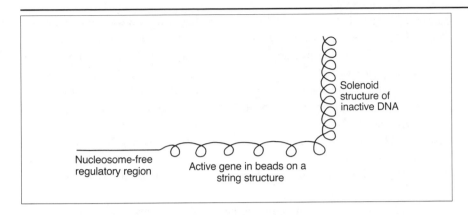

Figure 5.31 Levels of chromatin structure in active and inactive DNA.

Solenoid structure of inactive DNA

Nucleosome-free regulatory region

Active gene in beads on a string structure

entirely nucleosome-free DNA which may have an unusual conformation.

The role of these changes in allowing cells to maintain a commitment to a particular differentiated state and to respond differently to inducers of gene expression is well illustrated in the case of the steroid hormones and their effect on gene expression. Thus the difference in the response of different tissues to treatment with estrogen (Section 5.1) is likely to be due to the fact that in one tissue certain steroid-responsive genes will be inaccessible within the solenoid structure and will therefore be incapable of binding the receptor–hormone complex that is necessary for activation. In other genes, which are in the beads on a string structure and therefore more accessible, such binding of the complex to defined sequences in the gene will occur. Even in this case, however, gene activation will not occur as a direct consequence of this interaction as might be the case in bacteria. Rather, the binding will result in the displacement of a nucleosome from the region of DNA, generating a hypersensitive site and allowing other regulatory proteins to interact with their specific recognition sequences and cause transcription to occur.

Both the action of the receptor–hormone complex and the subsequent binding of other transcription factors to nucleosome-free DNA clearly involve the interaction of regulatory proteins with specific DNA sequences. The next two chapters will discuss the DNA sequences and proteins involved in these interactions.

References

Ball, D. J., D. S. Gross and W. T. Garrard 1983. 5-methyl cytosine is localized in nucleosomes that contain H1. *Proceedings of the National Academy of Sciences of the USA* **80**, 5490–4.

Ballabio, A. and H. F. Willard 1992. Mammalian X chromosome inactivation and the XIST gene. *Current Opinion in Genetics and Development* **2**, 439–47.

Barlow, D. P. 1993. Methylation and imprinting: from host defense to gene regulation? *Science* **260**, 309–10.

Beato, M. 1989. Gene regulation by steroid hormones. *Cell* **56**, 335–44.

Becker, P. B., S. Ruppert and G. Schutz 1987. Genomic fingerprinting reveals cell type-specific binding of ubiquitous factors. *Cell* **51**, 435–43.

Bestor, T. H. and A. Coxon 1993. The pros and cons of DNA methylation. *Current Biology* **3**, 384–6.

Bird, A. 1984. DNA methylation – how important in gene control? *Nature* **307**, 503.

Bird, A. 1987. CpG islands as gene markers in the vertebrate nucleus. *Trends in Genetics* **3**, 342–7.

Bird, A. 1992. The essentials of DNA methylation. *Cell* **70**, 5–8.

Boyes, J. and A. Bird 1991. DNA methylation inhibits transcription indirectly via a methyl-CpG binding protein. *Cell* **64**, 1123–34.

Busch, H. and I. L. Goldknopf 1981. Ubiquitin–protein conjugates. *Molecular and Cellular Biochemistry* **40**, 173–87.

Cedar, H. 1988. DNA methylation and gene activity. *Cell* **53**, 3–4.

Chiu, C.-P. and H. Blau 1985. 5-azacytidine permits gene activation in a previously non-inducible cell type. *Cell* **40**, 417–24.

Constantinides, P. G., P. A. Jones and W. Gevers 1977. Functional striated muscle cells from non-myoblast precursors following 5-azacytidine treatment. *Nature* **267**, 364–6.

Coon, H. G. 1966. Clonal stability and phenotypic expression of chick cartilage cells *in vitro*. *Proceedings of the National Academy of Sciences of the USA* **55**, 66–73.

Elgin, S. C. R. 1984. Anatomy of hypersensitive sites. *Nature* **309**, 213–14.

Felsenfeld, G. 1992. Chromatin as an essential part of the transcriptional mechanism. *Nature* **355**, 219–24.

Felsenfeld, G. and J. D. McGhee 1986. Structure of the 30 nm chromatin fiber. *Cell* **44**, 375–7.

Finch, J. T., M. Noll and R. D. Kornberg 1975. Electron microscopy of defined lengths of chromatin. *Proceedings of the National Academy of Sciences of the USA* **72**, 3320–2.

Gottesman, S. 1984. Bacterial regulation: global regulatory networks. *Annual Review of Genetics* **18**, 415–41.

Grant, S. G. and V. M. Chapman 1988. Mechanisms of X chromosome regulation. *Annual Review of Genetics* **22**, 199–233.

Gross, D. S. and W. T. Garrard 1988. Nuclease hypersensitive sites in chromatin. *Annual Review of Biochemistry* **57**, 159–97.

Hadorn, E. 1968. Transdetermination in cells. *Scientific American* **219** (November), 110–20.

Hayashizaki, Y. *et al.* 1994. Identification of an imprinted U2af binding protein related sequence on mouse chromosome 11 using the RLGS method. *Nature Genetics* **6**, 33–40.

Igo-Kemenes, T., W. Horz and H. G. Zachau 1982. Chromatin. *Annual Review of Biochemistry* **51**, 89–121.

Kaye, J. S., M. Bellard, G. Dretzen, F. Bellard and P. Chambon 1984. A close association between sites of DNase I hypersensitivity and sites of enhanced cleavage by microccocal nuclease in the 5' flanking region of the actively transcribed ovalbumin gene. *EMBO Journal* **3**, 1137–44.

Keene, M. A., V. Corces, K. Lowenhaupt and S. Elgin 1981. DNase I hypersensitive sites in *Drosophila* chromatin occur at the 5' end of regions of transcription. *Proceedings of the National Academy of Sciences of the USA* **78**, 143–6.

Keshet, I., J. Lieman-Hurwitz and H. Cedar 1986. DNA methylation affects the formation of active chromatin. *Cell* **44**, 535–44.

Kornberg, R. D. and A. Klug 1981. The nucleosome. *Scientific American* **244** (February), 48–64.

Kumar, S., X. Cheng, S. Klimasauskas, S. Mi, J. Posfai, R. J. Roberts and G. G. Wilson 1994. The DNA (cytosine-5) methyltransferases. *Nucleic Acids Research* **22**, 1–10.

Kuroda, M. I., M. J. Palmer and J. C. Lucchesi 1993. X chromosome dosage compensation in *Drosophila*. *Seminars in Developmental Biology* **4**, 107–16.

Lacey, E. and R. Axel 1975. Analysis of DNA of isolated chromatin subunits. *Proceedings of the National Academy of Sciences of the USA* **72**, 3978–82.

Levinger, L. and A. Varshavsky 1982. Selective arrangement of ubiquitinated and D1 protein containing nucleosomes within the *Drosophila* genome. *Cell* **28**, 375–85.

Lewin, B. 1980. *Gene Expression*. Vol. 2: *Eukaryotic Chromosomes*. Wiley, New York.

Li, E., T. H. Bestor and R. Jaenisch 1992. Targeted mutation of the DNA methyl-transferase gene results in embryonic lethality. *Cell* **69**, 915–26.

Lorch, Y., J. W. La Pointe and R. D. Kornberg 1987. Nucleosomes inhibit the initiation of transcription but allow chain elongation with the displacement of histones. *Cell* **49**, 203–10.

Lyon, M. F. 1992. Some milestones in the history of X chromosome inactivation. *Annual Review of Genetics* **26**, 17–28.

Lyon, M. F. 1993. Epigenetic inheritance in mammals. *Trends in Genetics* **9**, 123–8.

McGhee, J. D., W. I. Wood, M. Dolan, J. D. Engel and G. Felsenfeld 1981. A 200 base pair region at the 5' end of the chicken adult β-globin gene is accessible to nuclease digestion. *Cell* **27**, 45–55.

McKeon, C., H. Ohkubo, I. Pastan and B. de Crombrugghe 1982. Unusual methylation pattern of the alpha 2 (1) collagen gene. *Cell* **29**, 203–10.

McKnight, S. L., M. Bustin and O. L. Miller 1978. Electron microscope analysis of chromosome metabolism in the *Drosophila melanogaster* embryo. *Cold Spring Harbor Symposium on Quantitative Biology* **42**, 741–54.

Miller, D. M., P. Turner, A. W. Nienhuis, D. E. Axelrod and T. U. Gopalakrishnan 1978. Active conformation of the globin genes in uninduced and induced mouse erythroleukemia cells. *Cell* **14**, 511–21.

Miller, J. and W. K. Reznikoff (eds) 1980. *The Operon.* Cold Spring Harbor Laboratory Press, Cold Spring Harbor, NY.

Moore, T. and D. Haig 1991. Genomic imprinting in mammalian development. *Trends in Genetics* **7**, 45–9.

Morse, R. H. and R. T. Simpson 1988. DNA in the nucleosome. *Cell* **54**, 285–7.

Nahon, J.-L., A. Venetianer and J. M. Sala-Trepat 1987. Specific sets of DNase I hypersensitive sites are associated with the potential and overt expression of the rat albumin and alpha-fetoprotein genes. *Proceedings of the National Academy of Sciences of the USA* **84**, 2135–9.

Pardue, M. L., A. Nordheim, E. M. Lafer, B. D. Stollar and A. Rich 1983. ZDNA and the polytene chromosome. *Cold Spring Harbor Symposia on Quantitative Biology* **47**, 171–6.

Raibaud, O. and M. Schwartz 1984. Positive control of transcription initiation in bacteria. *Annual Review of Genetics* **18**, 173–206.

Razin, A. and H. Cedar 1991. DNA methylation and gene expression. *Microbiological Reviews* **55**, 451–8.

Reeves, R. 1984. Transcriptionally active chromatin. *Biochimica et Biophysica Acta* **782**, 343–93.

Reeves, R. and P. Cserjesi 1979. Sodium butyrate induces new gene expression in Friend erythroleukemic cells. *Journal of Biological Chemistry* **254**, 4283–90.

Reik, W. and N. D. Allen 1994. Imprinting with and without methylation. *Current Biology* **4**, 145–7.

Rich, A., A. Nordheim and A. H.-J. Wang 1984. The chemistry and biology of left-handed DNA. *Annual Review of Biochemistry* **53**, 791–846.

Roth, S. Y and C. D. Allis 1992. Chromatin condensation: does histone H1 dephosphorylation play a role? *Trends in Biochemical Sciences* **17**, 93–8.

Saragosti, S., G. M. Moyne and M. Yaniv 1980. Absence of nucleosomes in a fraction of SV40 chromatin between the origin of replication and the region coding for the late leader RNA. *Cell* **20**, 65–73.

Shermoen, A. W. and S. K. Beckendorf 1982. A complex of interacting DNAse I hypersensitive sites near the *Drosophila* glue protein gene sgs4. *Cell* **29**, 601–7.

Stalder, J., M. Groudine, J. B. Dodgson, J. D. Engel and H. Weintraub 1980a. Hb switching in chickens. *Cell* **19**, 973–80.

Stalder, J., A. Larsen, J. D. Engel, M. Dolan, M. Groudine and H. Weintraub 1980b. Tissue-specific DNA cleavages in the globin chromatin domain introduced by DNAse I. *Cell* **20**, 451–60.

Surani, M. A. 1991. Genomic imprinting: developmental significance and molecular mechanism. *Current Opinion in Genetics and Development* **1**, 241–6.

Surani, M. A. 1993. Silence of the genes. *Nature* **366**, 302–3.

Thoma, F., T. Koller and A. Klug 1979. Involvement of histone H1 in the organization of the nucleosome and of the salt-dependent superstructures of chromatin. *Journal of Cell Biology* **83**, 403–27.

Travers, A. 1993. *DNA–Protein Interactions.* Chapman & Hall, London.

Turner, B. M. 1993. Decoding the nucleosome. *Cell* **75**, 5–8.

Uriel-Shoval, S., Y. Grunebaum, J. Sedat and A. Razin 1982. The absence of detectable methylated bases in *Drosophila melanogaster* DNA. *FEBS Letters* **146**, 148–52.

Weintraub, H. 1985. Assembly and propagation of repressed and derepressed chromatin states. *Cell* **42**, 705–11.

Weintraub, H., A. Larsen and M. Groudine 1981. Alpha globin gene switching during the development of chicken embryos: expression and chromosome structure. *Cell* **24**, 333–44.

Weisbrod, S. 1982. Active chromatin. *Nature* **297**, 289–95.

Weisbrod, S. and H. Weintraub 1979. Isolation of a sub-class of nuclear proteins responsible for conferring a DNAse I-sensitive structure on globin chromatin. *Proceedings of the National Academy of Sciences of the USA* **76**, 630–4.

Workman, J. L. and A. R. Buchman 1993. Multiple functions of nucleosomes and regulatory factors in transcription. *Trends in Biochemical Sciences* **18**, 90–5.

6 Transcriptional control – DNA sequence elements

6.1 Introduction

6.1.1 Relationship of gene regulation in prokaryotes and eukaryotes

As discussed in Chapter 5, various alterations occur in the chromatin structure of a particular gene prior to the onset of transcription. Once such changes have occurred, the actual onset of transcription takes place through the interaction of defined proteins (transcription factors) with specific DNA sequences adjacent to the gene. This final stage of gene regulation is clearly analogous to the activation or repression of gene expression in prokaryotes, which was discussed briefly in Chapter 5 (Section 5.1). However, before the manner in which transcription factors and DNA sequences act to regulate gene expression in higher eukaryotes can be considered, it is necessary to mention two features of eukaryotic systems that do not exist in bacteria.

6.1.2 Complexity of the eukaryotic system

RNA polymerases
In prokaryotes a single RNA polymerase enzyme is responsible for the transcription of DNA into RNA. In eukaryotes this is not the case, and three such enzymes, active on distinct sets of genes, exist and can be distinguished by their relative sensitivity to the fungal toxin α-amanitin (Table 6.1). Thus, whereas all genes capable of encoding a protein, as well as the genes for some small nuclear RNAs involved in RNA splicing (Section 4.2), are transcribed by RNA polymerase II, the genes encoding the 28S, 18S and 5.8S ribosomal RNAs are transcribed by RNA polymerase I, and those encoding the transfer RNAs and the 5S ribosomal RNA are transcribed by RNA polymerase III (reviewed by Sentenac, 1985).

Table 6.1 Eukaryotic RNA polymerases

	Genes transcribed	Sensitivity to α-amanitin
I	Ribosomal RNA (45S precursor of 28S, 18S and 5.8S rRNA)	Insensitive
II	All protein-coding genes, small nuclear RNAs U1, U2, U3, etc.	Very sensitive (inhibited 1 μg/ml)
III	Transfer RNA, 5.8S ribosomal RNA, small nuclear RNA U6, repeated DNA sequences: Alu, B1, B2, etc.; 7SK, 7SL RNA	Moderately sensitive (inhibited 10 μg/ml)

In considering the transcriptional regulatory processes that produce tissue-specific variation in mRNA and protein, our primary concern will therefore be with the regulation of RNA polymerase II transcription. The DNA sequence elements involved in regulating transcription by this enzyme are considered in Sections 6.2–6.4 of this chapter and the proteins that bind to them are discussed in Chapter 7. Transcription by RNA polymerases I and III is also subject to regulation, however, and has been analyzed in particular detail in the case of the transcription of the genes encoding the oocyte-specific form of 5S ribosomal RNA by RNA polymerase III. The processes regulating transcription by these polymerases are distinct from those operating on RNA polymerase II, and are considered in a separate section of this chapter (Section 6.5).

Co-ordinately regulated genes are not linked in eukaryotes

Even when considering transcription by RNA polymerase II alone, eukaryotic genes exhibit one further complication when compared with prokaryotes. In prokaryotes, when several genes encoding particular proteins are expressed in response to a particular signal, the genes are found tightly linked together in an operon. In response to the activating signal, all the genes in the operon are transcribed as one single poly-cistronic (multi-gene) mRNA molecule and translation of this molecule results in the desired co-ordinate production of the proteins encoded by the individual genes. A typical example of this is seen in the case of the lactose-inducible genes located in the *lac* operon. The efficient use of lactose requires not only the enzyme β-galactosidase, which cleaves the lactose molecule, but also a permease enzyme, to facilitate uptake of lactose by the cell, and a modifying *trans*-acetylase enzyme. The genes encoding these three molecules must be activated in response to lactose and this is achieved by linking them together in an operon whose expression is dependent upon the presence of lactose (Figure 6.1).

Figure 6.1 Structure of the *lac* operon of *Escherichia coli*, in which the three genes are transcribed into one single RNA and translated into individual proteins following binding of the ribosome at three sites in the RNA molecule.

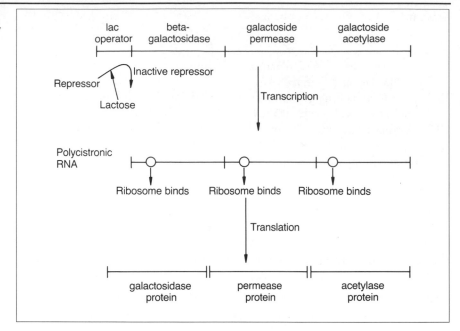

In higher eukaryotic organisms this does not occur. Thus operon-type structures transcribed into polycistronic RNAs do not exist in higher organisms. Rather, individual genes are transcribed into individual mono-cistronic RNAs encoding single proteins. Hence, genes whose protein products are required in parallel are present in the genome as individual genes whose expression must be co-ordinated. Moreover, such co-ordinately expressed genes are not even closely associated in the genome in a manner that might be thought to facilitate their regulation, but are very often present on different chromosomes within the eukaryotic nucleus. Thus, the production of a functional antibody molecule by the mammalian B cell requires the synthesis of both the immunoglobulin heavy- and light-chain proteins, which together make up the functional antibody molecule. The genes encoding the heavy- and light-chain proteins are, however, found on separate chromosomes; the heavy-chain locus being on chromosome 14 in humans whereas light-chain genes are found on both chromosomes 2 and 22. Similarly, the production of a functional globin molecule requires the association of an α-globin-type protein and a β-globin-type protein. Yet the genes encoding the various members of the α-globin family are on chromosome 16 in humans whereas the genes encoding the β-globins are found on chromosome 11 (for review of the globin gene family, see Evans *et al.*, 1990).

This difference between prokaryotes and eukaryotes is likely to reflect a need for greater flexibility in regulating gene expression in eukaryotes.

Thus the bacterial arrangement, although providing a simple mechanism for co-ordinating the expression of different proteins, also means that, in general, expression of one protein will also necessitate expression of the other co-ordinately expressed proteins. In contrast, the eukaryotic system allows α-globin to be produced in parallel with β-globin in the adult organism but to be associated with another β-globin-like protein, i.e. γ-globin, in the fetus.

6.1.3   The Britten and Davidson model for the co-ordinate regulation of unlinked genes

This greater flexibility does, however, necessitate some means of co-ordinately regulating gene expression, allowing the production of α- and β-globin in the adult reticulocyte, α- and β-globin in the embryonic reticulocyte, and the heavy and light chains of immunoglobulin in the antibody-producing B cell. A model of the mechanism of such co-ordinate regulation was put forward by Britten and Davidson (1969). They proposed that genes regulated in parallel with one another in response to a particular signal would contain a common regulatory element which would cause the activation of the gene in response to that signal (Figure 6.2). Individual genes could contain more than one regulatory element, some of which would be shared with other genes, which in turn could possess elements not present in the first gene. Specific signals causing gene activation would act by stimulating a specific integrator gene whose product would activate all the genes containing one particular sequence element. This mechanism would allow the observed activation of distinct but overlapping sets of genes in response to specific signals via the activation of particular integrator genes.

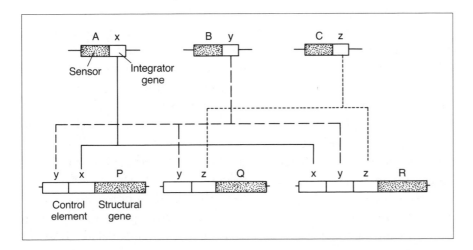

Figure 6.2 Britten and Davidson model for the co-ordinate expression of unlinked genes in eukaryotes. Sensor elements (A, B, C) detect changes requiring alterations in gene expression and switch on appropriate integrator genes (x, y, z) whose products activate the structural genes (P, Q, R) containing the appropriate control elements. Note the flexibility of the system whereby a particular structural gene can be activated with or without another structural gene by selecting which integrator gene is activated.

Although this model was proposed over 20 years ago, when our understanding of eukaryotic gene regulation was considerably more limited than at present, it continues to serve as a useful framework for considering gene regulation. In modern terms, the integrator gene would be considered as encoding a transcription factor (Chapter 7) which binds to the regulatory sequences and activates expression of the corresponding genes. As discussed in Chapter 7 (Section 7.4), the activation of such a factor by a particular signal has now been shown to operate both by *de novo* synthesis of the factor (in an individual tissue or in response to a particular signal) as envisaged by Britten and Davidson, and by direct activation of a pre-existing protein, e.g. by phosphorylation in response to the activating signal. The concept that such factors act by binding to DNA sequences held in common between co-ordinately regulated genes is now well established, and it is therefore necessary to consider the nature of these sequences. In the original model of Britten and Davidson (1969) it was considered that this role would be played by a class of repeated DNA sequences about 200–300 bp in length which had been shown to be present in many copies in the genome (for review, see Jelinek and Schmid, 1982). It is now clear, however, that the target DNA binding sites for transcription factors are much shorter sequences approximately 10–30 bp in length and these short sequence elements will now be discussed (for review, see Maniatis *et al.*, 1987).

6.2 Short sequence elements located within or adjacent to the gene promoter

6.2.1 Short regulatory elements

In prokaryotes the sequence elements that play an essential role in transcriptional regulation are located immediately upstream of the point at which transcription begins and form part of the promoter which directs transcription of the gene. Such sequences can be divided into two classes, i.e. those found in all genes, which play an essential role in the process of transcription itself, and those found in one or a few genes, which mediate their response to a particular signal (Schmitz and Galas, 1979). It would be expected by analogy, therefore, that the sequences that play a role in the regulation of eukaryotic transcription would be located similarly, upstream of the transcription start site within or adjacent to the promoter. Hence, a comparison of such sequences in different genes should reveal both basic promoter elements necessary for all transcription (which would be present in all genes) and those necessary for a particular pattern of regulation (which would be present only in genes exhibiting a specific pattern of regulation). The role of such sequences can be confirmed either by destroying them by deletion or mutation or by transferring them to other genes in an attempt to confer the specific pattern of regulation of the donor gene upon the recipient.

6.2.2 The 70 kDa heat-shock protein gene

To illustrate this method of analysis, we will focus upon the gene encoding the 70 kDa heat-shock protein (hsp70) and compare it with other genes showing different patterns of regulation. As discussed in Chapter 3 (Section 3.2.4), exposure of a very wide variety of cells to elevated temperature results in the increased synthesis of a few heat-shock proteins, of which hsp70 is the most abundant. Such increased synthesis is mediated in part by increased transcription of the corresponding gene, which can be visualized as a puff within the polytene chromosome of *Drosophila* (for review, see Ashburner and Bonner, 1979). Hence, examination of the sequences located upstream of the start site for transcription in this gene should identify potential sequences involved in its induction by temperature elevation, as well as those involved in the general mechanism of transcription. The sequences present in this region of the *hsp70* gene which are also found in other genes are listed in Table 6.2, and their arrangement is illustrated in Figure 6.3 (Williams *et al.*, 1989; for further information on these sequences see Davidson *et al.*, 1983; Jones *et al.*, 1988 and references therein).

A comparison of this type reveals a number of sequence motifs shared by the *hsp70* gene and other non-heat-inducible genes as well as one which is unique to heat-inducible genes. These will be considered in turn.

The TATA box

The TATA box is an AT-rich sequence which is found approximately 30 bases upstream of the transcription start site in a wide variety of different genes, although it is absent from some housekeeping genes expressed in all tissues and a few tissue-specific genes. In genes that contain it, the TATA

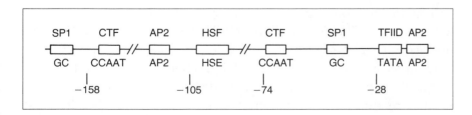

Figure 6.3 Transcriptional control elements in the human *hsp70* gene promoter. The protein binding to a particular site is indicated above the line and the corresponding DNA element below the line. These elements are described more fully in Table 6.2.

Table 6.2   Sequences present in the upstream region of the *hsp70* gene which are also found in other genes

Name	Consensus	Other genes containing sequences
TATA box	TATA A/T A A/T	Very many genes
CCAAT box	TGTGGCTNNNAGCCAA	α- and β-globin, herpes simplex virus thymidine kinase, cellular oncogenes c-*ras*, c-*myc*, albumin, etc.
Sp1 box	GGGCGG	Metallothionein II A, type II procollagen, dihydrofolate reductase, etc.
CRE	T/G T/A CGTCA	Somatostatin, fibronectin, α-gonadotrophin, c-*fos*, etc.
AP2 box	CCCCAGGC	collagenase, class 1 antigen H-2K^b, metallothionein II A
Heat-shock consensus	CTNGAATNTTCTAGA	Heat-inducible genes *hsp83*, *hsp27*, etc.

box plays a critical role in positioning the start site of transcription and its destruction by mutation or deletion effectively abolishes transcription of such genes (for review, see Breathnach and Chambon, 1981).

The TATA box is bound by a complex of proteins known as TFIID (transcription factor D for RNA polymerase II). Within this complex only one protein, known as TBP (TATA-binding protein; for review, see Hernandez, 1993), actually binds directly to the DNA, with the other proteins in the TFIID complex binding via their association with TBP. The binding of TBP to the TATA box together with its associated proteins is essential for transcription to occur since such binding allows the subsequent binding of other factors such as TFIIB and of the RNA polymerase itself to form a basal transcriptional complex (Figure 6.4; for reviews, see Roeder, 1991; Zawel and Reinberg, 1992; Buratowski, 1994). All these other factors, including the RNA polymerase, bind via protein–protein interactions with already bound proteins rather than by recognizing specific DNA sequences. The binding of TBP to the TATA box is therefore essential for transcription.

Interestingly, a number of genes transcribed by RNA polymerase II do not contain a TATA box but still require TBP for transcription (for review, see Weis and Reinberg, 1992). In this case it is likely that another factor binds to the promoter DNA first and TBP then binds via a protein–protein interaction (Figure 6.5). This indicates that TBP is likely to play an essential role in the transcription of all promoters transcribed by RNA

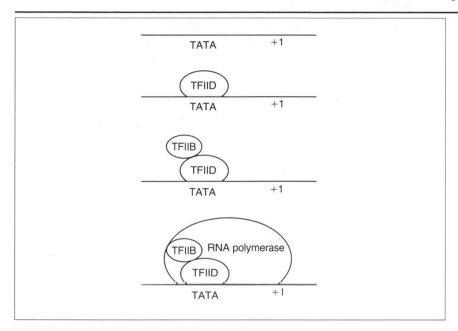

Figure 6.4 Formation of the basal transcriptional complex involves the initial binding of TFIID (TBP and associated proteins) to the TATA box which is followed by the binding of other components such as TFIIB and the RNA polymerase itself.

polymerase II, perhaps by interacting directly with the RNA polymerase itself to stimulate its activity (Figure 6.5a and b). In addition in TATA-box-containing promoters it binds to the TATA box, allowing other factors to bind subsequently via protein–protein interactions (for review, see Hernandez, 1993).

The CCAAT and Spl boxes

The basal transcriptional complex containing TBP, RNA polymerase II and other associated factors can produce only a low rate of transcription. This rate is enhanced by the binding of other transcription factors to sites upstream of the TATA box which enhances either the stability or the activity of the basal complex. The binding sites for several of these factors

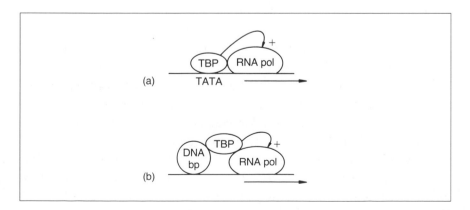

Figure 6.5 Although TBP can bind directly to the TATA box in promoters which contain it (a), it binds via a protein–protein interaction with another DNA-binding protein (DNA bp) in promoters which lack the TATA box (b).

are present in a wide variety of different genes with different patterns of activity. These sites act as targets for the binding of specific factors which are active in all cell types and their binding therefore results in increased transcription in all tissues (for review, see Latchman, 1991).

An example of this is the Spl box, two copies of which are present in the hsp70 gene promoter (Figure 6.3). This GC-rich DNA sequence binds a transcription factor known as Spl which is present in all cell types (for review, see McKnight and Tjian, 1986). Similarly, the CCAAT box, which is located upstream of the start site of transcription of a wide variety of genes (including the *hsp70* gene) which are regulated in different ways, is also believed to play an important role in allowing transcription of the genes containing it by binding constitutively expressed transcription factors (for review, see McKnight and Tjian, 1986).

The heat-shock element

In contrast to these very widespread sequence motifs, another sequence element in the *hsp70* gene is shared only with other genes whose transcription is increased in response to elevated temperature. This sequence is found 62 bases upstream of the start site for transcription of the *Drosophila hsp70* gene and at a similar position in other heat-inducible genes (Davidson *et al.*, 1983). This heat-shock consensus element is therefore believed to play a critical role in mediating the observed heat inducibility of transcription of these genes. In order to confirm that this is the case, it is necessary to transfer this sequence from the *hsp70* gene to another gene, which is not normally heat inducible, and show that the recipient gene now becomes inducible. This was achieved by Pelham (1982) who transferred the heat-shock consensus element onto the non-heat-inducible thymidine kinase (*tk*) gene taken from the eukaryotic virus, herpes simplex. When the hybrid gene was introduced into cells and the temperature subsequently raised, increased thymidine kinase production was detected, showing that the heat-shock consensus element had rendered the *tk* gene inducible by elevated temperature (Figure 6.6).

This experiment therefore proves that the common sequence element found in the heat-inducible genes is responsible directly for their heat inducibility. The manner in which these experiments were carried out also permits a further conclusion with regard to the way in which this sequence acts. Thus, the heat-shock sequence element used by Pelham was taken from the *Drosophila hsp70* gene and, in this cold-blooded organism, would be activated normally by the thermally stressful temperature of 37°C. The cells into which the hybrid gene was introduced, however, were mammalian cells which grow normally at 37°C and only express the heat-

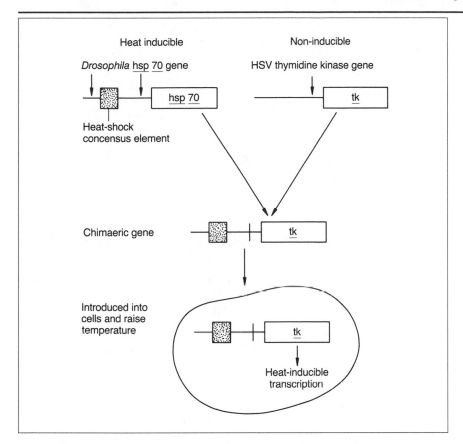

Figure 6.6 Demonstration that the heat-shock consensus element mediates heat inducibility. Transfer of this sequence to a gene (thymidine kinase) which is not normally inducible renders this gene heat inducible.

shock genes at the higher temperature of 42°C. In these experiments the hybrid gene was induced only at 42°C, the heat-shock temperature characteristic of the cell into which it was introduced, and not at 37°C, the temperature characteristic of the species from which the DNA sequence came. This means that the heat-shock consensus sequence does not possess some form of inherent temperature sensor or thermostat which is set to go off at a particular temperature, since in this case the *Drosophila* sequence would activate transcription at 37°C, even in mammalian cells. Rather, it must act by being recognized by a cellular protein which is activated in response to elevated temperature and, by binding to the heat-shock element, produces increased transcription. Evidently, although the elements of this response are conserved sufficiently to allow the mammalian protein to recognize the *Drosophila* sequence, the mammalian protein will, of course, only be activated at the mammalian heat-shock temperature and hence induction will only occur at 42°C.

Hence these experiments not only provide evidence for the importance of the heat-shock consensus element in causing heat-inducible transcription, but also indicate that it acts by binding a protein. Direct evidence

Figure 6.7 Detection of a protein binding to a DNA sequence by inhibition of DNA digestion with exonuclease III.

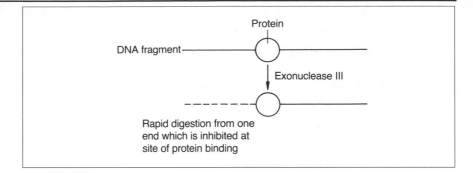

that this is the case was provided by using techniques such as digestion of the DNA with the enzyme exonuclease III. Although this enzyme will progressively digest DNA, starting at one end, its progress is impeded by the presence of a protein on the DNA, and hence the positions of such proteins can be mapped (Figure 6.7). When an analysis of this type is carried out on the upstream regions of the heat-shock genes, it can be shown that in non-heat-shocked cells the TFIID factor is bound to the TATA box and an additional factor known as the GAGA factor is bound to upstream sequences (Figure 6.8a). The binding of the GAGA factor is believed to displace a nucleosome and create the DNaseI-hypersensitive sites observed in these genes in non-heat-shocked cells (Chapter 5) (for review, see van Holde, 1994). Hence the potentially activatable state illustrated in Figure 5.27(a) is created by the binding of the GAGA factor.

By contrast, in heat-shocked cells, where high-level transcription of the gene is occurring, an additional protein, which is bound to the heat-shock consensus element, is detectable on the upstream region (Figure 6.8b). Hence the induction of the heat-shock genes is indeed accompanied by the binding of a protein, known as the heat-shock factor (HSF), to the

Figure 6.8 Proteins binding to the promoter of the *hsp70* gene before (a) and after (b) heat shock.

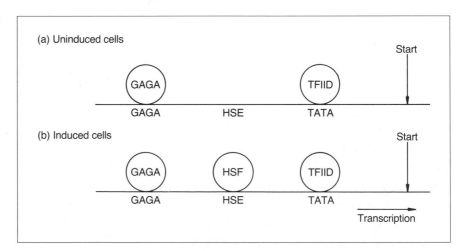

heat-shock consensus sequence, as suggested by the experiments of Pelham (1982). The binding of this factor to a gene whose chromatin structure has already been altered to render it potentially activatable results in stimulation of transcription exactly as discussed in Chapter 5 and illustrated in Figure 5.27. In agreement with this, the purified heat-shock factor can bind to the heat-shock consensus element and stimulate the transcription of the *hsp70* gene in a cell-free nuclear extract while having no effect on the transcription of the non-heat-inducible actin gene (Parker and Topol, 1984).

The activation of these heat-inducible genes can therefore be fitted very readily into the Britten and Davidson scheme for the regulation of gene expression. The heat-shock consensus sequence represents the common sequence present in the similarly regulated genes, whereas the heat-shock factor would represent the product of the integrator gene which regulates their expression. Unlike in the original model, however, it is clear that the heat-shock factor is not synthesized *de novo* in response to thermal stress. Rather it is present in unstressed cells (Parker and Topol, 1984) and can activate the heat-shock genes following exposure to elevated temperature even in the presence of protein synthesis inhibitors preventing its *de novo* synthesis (Zimarino and Wu, 1987). It is clear, therefore, that upon heat shock a previously inactive factor is activated to a DNA-binding form by a post-translational modification involving an alteration of the protein itself. This activation has been produced in an isolated cell-free nuclear extract by elevated temperature, and appears to involve a temperature-dependent change in the structure of the protein which allows it to bind to the heat-shock element and activate transcription (for review, see Morimoto, 1993).

6.2.3 Other response elements

Although some modification of the Britten and Davidson model, with regard to the activation of the transcription factor itself, is therefore necessary, the heat-shock system does provide a clear example of the role of short common sequences within the promoters of particular genes in mediating their common response to a specific stimulus. A number of similar elements, which are found in the promoters of genes activated by other signals, have now been identified and have been shown to be capable of transferring the specific response to another marker gene (reviewed by Davidson *et al.*, 1983). A selection of such sequences is listed in Table 6.3.

As indicated in Table 6.3, these sequences act by binding specific proteins which are synthesized or activated in response to the inducing

Table 6.3 Sequences that confer response to a particular stimulus

Consensus sequence	Response to	Protein factor	Genes containing sequences
CTNGAATNTT CTAGA	Heat	Heat-shock transcription factor	*hsp70, hsp83, hsp27*, etc.
T/G T/A CGTCA	Cyclic AMP	CREB/ATF	Somatostatin, fibronectin, α-gonadotrophin, c-*fos*, *hsp70*
TGAGTCAG	Phorbol esters	AP1	Metallothionein II A, α-1-antitrypsin, collagenase
GATGTCCATATT AGGACATC	Growth factors in serum	Serum response factor	c-*fos*, *Xenopus* γ-actin
GGTACANNN TGTTCT	Glucocorticoid, progesterone	GR and PR receptors	Metallothionein II A, tryptophan oxygenase, uteroglobin, lysozyme
AGGTCANNN TGACCT	Estrogen	Estrogen receptor	Ovalbumin, conalbumin, vitellogenin
TCAGGTCAT GACCTGA	Thyroid hormone, retinoic acid	TH and RA receptors	Growth hormone, myosin heavy chain
TGCGCCCGCC	Heavy metals	Not known	Metallothionein genes
AAGTGA	Viral infection	Not known	Interferon-α and -β, tumor necrosis factor

signal. Such transcription factors are discussed further in Chapter 7. It is noteworthy, however, that many of the sequences in Table 6.3 exhibit dyad symmetry, a similar sequence being found in the 5' to 3' direction on each strand. The estrogen response element, for example, has the sequence:

$$5'\text{-A G G T C A N N N T G A C C T-}3'$$
$$3'\text{ T C C A G T N N N A C T G G A-}5'$$

The two halves of the 10 base palindrome are separated by three random bases. Such symmetry in the binding sites for these transcription factors suggests that they may bind to the site in a dimeric form consisting of two protein molecules.

Various sequences that confer response to several different signals have therefore been identified. Exactly as suggested by Britten and Davidson, one gene can possess more than one such element, allowing multiple patterns of regulation. Thus comparison of the sequences listed in Table 6.3 with those contained in the *hsp70* gene, listed in Table 6.2, reveals that, in addition to the heat-shock consensus element, this gene also contains the cyclic AMP response element (CRE) which mediates the induction of a number of genes, such as that encoding somatostatin, in response to treatment with cyclic AMP. Similarly, while genes may share particular elements, flexibility is provided by the presence of other elements in one gene and not another, allowing the induction of a particular gene in response to a given stimulus which has no effect on another gene. For example, although the *hsp70* gene and the metallothionein IIA gene share a binding site for the transcription factor AP2, only the metallothionein gene has a binding site for the glucocorticoid receptor, which confers responsivity to glucocorticoid hormone induction, and hence only this gene is inducible by hormone treatment.

In some cases, the sequence elements that confer response to a particular stimulus can be shown to be related to one another. Thus the sequences mediating response to glucocorticoid or progesterone treatment are similar to that which mediates response to another steroid, i.e. estrogen. Similarly, both the estrogen- and thyroid hormone-responsive elements contain identical sequences showing dyad symmetry. In the estrogen-responsive element, however, the two halves of this dyad symmetry are separated by three bases which vary between different genes, whereas in the thyroid hormone-responsive element the two halves are contiguous (Table 6.4; see Beato, 1989 for review). Such similarities are paralleled by a similarity in the individual cytoplasmic steroid receptor proteins which form a complex with each of these hormones and then bind to the corresponding DNA sequence. All of these receptors can be shown to be members of a large family of related DNA-binding proteins whose hormone and DNA-binding specificities differ from one another. The exchange of particular regions of these proteins with those of other family members has provided considerable information on the manner in which sequence-specific binding to DNA occurs and this is discussed in Chapter 7 (Section 7.2.3).

Table 6.4 Relationship of consensus sequences conferring responsivity to various hormones

Glucocorticoid/progesterone	GGTACANNNTGTTCT
Estrogen	AGGTCANNNTGACCT
Thyroid hormone/retinoic acid	TCAGGTCA——TGACCTGA

N indicates that any base can be present at this position; a dash indicates that no base is present, the gap having been introduced to align the sequence with the other sequences.

Although the sequence elements shown in Table 6.3 are all involved in the response to particular inducers of gene expression, it seems likely that other short sequence elements or combinations of elements will also be involved in controlling the tissue-specific patterns of expression exhibited by eukaryotic genes. Thus the octamer motif (ATGCAAAT), which is found in both the immunoglobulin heavy- and light-chain promoters, can confer B-cell-specific expression when linked to a non-regulated promoter (Wirth *et al.*, 1987). Similarly, short DNA sequences that bind liver-specific transcription factors have been identified in the region of the rat albumin promoter, known to be involved in mediating the liver-specific expression of this gene, and at least one of these sequence elements is also found in the liver-specific α-2 globulin gene (Lichtsteiner *et al.*, 1987).

6.2.4 Mechanism of action of promoter regulatory elements

It is clear, therefore, that short DNA sequence elements located near the start site of transcription play an important role in regulating gene expression in eukaryotes. As indicated in Table 6.3 and discussed above, such sequences mediate transcriptional activation by binding a specific protein. This binding may give rise to gene activity in one of two ways (Figure 6.9). First, as discussed in Chapter 5 and illustrated by the glucocorticoid receptor, binding of a specific protein may result in displacement of a nucleosome and generation of a DNaseI-hypersensitive site, allowing easy access to the gene for other transcription factors. The direct activation of transcription by such factors constitutes the second mechanism of gene induction, and is illustrated both by the binding of other non-regulated factors to glucocorticoid-regulated genes following binding of the receptor and by the binding of the heat-shock factor to its consensus sequence in the heat-inducible genes. These factors are likely to act by interacting directly with proteins necessary for transcription, such as TBP or RNA polymerase itself. This interaction facilitates the formation of a stable transcription complex, which may enhance the binding of RNA polymerase to the DNA or alter its structure in a manner which increases its activity (for further discussion, see Section 7.3.3; Latchman, 1991; Roeder, 1991).

It should be noted that these two mechanisms of action are not exclusive. Thus the glucocorticoid receptor, whose binding to a specific DNA sequence displaces a nucleosome, also contains an activation domain capable of interacting with other bound transcription factors (Section 7.3.2). Hence, following binding of the receptor, transcription is increased by inter-factor interactions between the receptor and other bound factors.

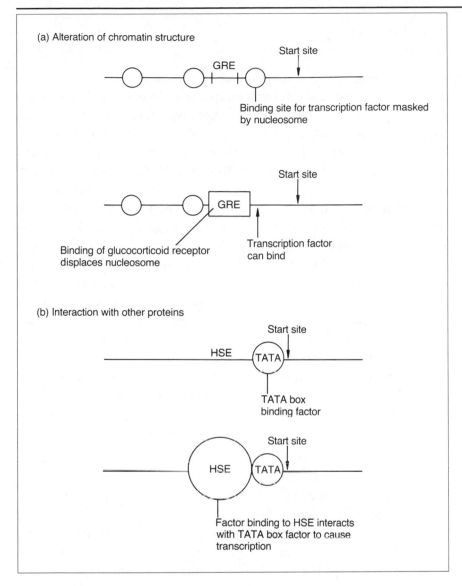

(a) Alteration of chromatin structure

Start site

GRE

Binding site for transcription factor masked
by nucleosome

Start site

GRE

Binding of glucocorticoid receptor
displaces nucleosome

Transcription factor
can bind

(b) Interaction with other proteins

Start site

HSE   TATA

TATA box
binding factor

Start site

HSE   TATA

Factor binding to HSE interacts
with TATA box factor to cause
transcription

Figure 6.9 Roles of short
sequence elements in gene
activation. These elements
can either bind a factor that
displaces a nucleosome and
unmasks a binding site for
another factor (a) or can bind
a factor that directly activates
transcription (b).

Short DNA sequence elements act in a similar manner to gene regulatory elements in prokaryotes and are located at a similar position close to
the start site of transcription. Unlike the situation in prokaryotes, however, important elements involved in the regulation of eukaryotic gene
expression are also found at very large distances from the site of initiation
of transcription. The nature of such regulatory elements will now be
discussed.

6.3   Enhancers

6.3.1   Regulatory sequences that act at a distance

The first indication that sequences located at a distance from the start site of transcription might influence gene expression in eukaryotes came with the demonstration that sequences over 100 bases upstream of the transcriptional start site of the histone H2A gene were essential for its high-level transcription (Grosschedl and Birnsteil, 1980). Moreover, although this sequence was unable to act as a promoter and direct transcription, it could increase initiation from an adjacent promoter element up to 100-fold when located in either orientation relative to the start site of transcription. Subsequently, a vast range of similar elements have been described in both cellular genes and those of eukaryotic viruses, and have been called enhancers because although they lack promoter activity and are unable to direct transcription themselves, they can dramatically enhance the activity of promoters (for reviews, see Hatzopoulos *et al.*, 1988; Muller *et al.*, 1988; Thompson and McKnight, 1992). Hence, if an enhancer element is linked to a promoter, such as that derived from the β-globin gene, the activity of the promoter can be increased several hundredfold. Variation in the position and orientation at which the enhancer element was placed relative to the promoter has led to three conclusions with regard to the action of enhancers. These are:

1. An enhancer element can activate a promoter when placed up to several thousand bases from the promoter.
2. An enhancer can activate a promoter when placed in either orientation relative to the promoter.
3. An enhancer can activate a promoter when placed upstream or downstream of the transcribed region, or within an intervening sequence which is removed from the RNA by splicing (Section 4.2).

These characteristics constitute the definition of an enhancer and are summarized in Figure 6.10.

6.3.2   Tissue-specific activity of enhancers

Following the discovery of enhancers, it was rapidly shown that many genes expressed in specific tissues also contained enhancers. Such enhancers frequently exhibited a tissue-specific activity, being able to enhance the activity of other promoters only in the tissue in which the gene from which they were derived is normally active and not in other tissues (Figure 6.11). The tissue-specific activity of such enhancers acting

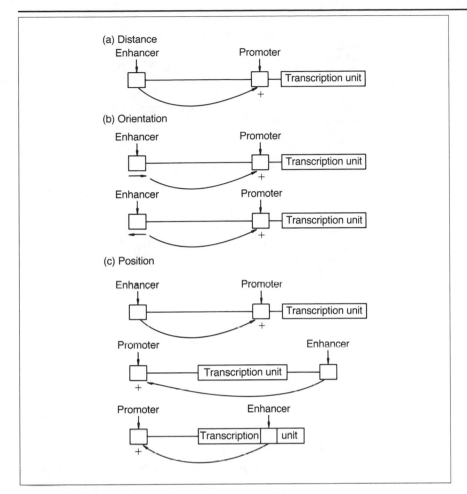

Figure 6.10 Characteristics of an enhancer element which can activate a promoter at a distance (a), in either orientation relative to the promoter (b) and when positioned upstream, downstream or within a transcription unit (c).

on their normal promoters is likely, therefore, to play a critical role in mediating the observed pattern of gene regulation.

Thus, as previously discussed (Section 2.4), the genes encoding the heavy and light chains of the antibody molecule contain an enhancer located within the large intervening region separating the regions encoding the joining and constant regions of these molecules. When this element is linked to another promoter, such as that of the β-globin gene, it increases its activity dramatically when the hybrid gene is introduced into B cells. In contrast, however, no effect of the enhancer on promoter activity is observed in other cell types, such as fibroblasts, indicating that the activity of the enhancer is tissue specific (Gillis *et al.*, 1983).

Similar tissue-specific enhancers have also been detected in genes expressed specifically in the liver (α-fetoprotein, albumin, α-1-anti-trypsin), the endocrine and exocrine cells of the pancreas (insulin, elastase, amylase), the pituitary gland (prolactin, growth hormone) and many

Figure 6.11 A tissue-specific enhancer can activate the promoter of its own or another gene only in one particular tissue and not in others.

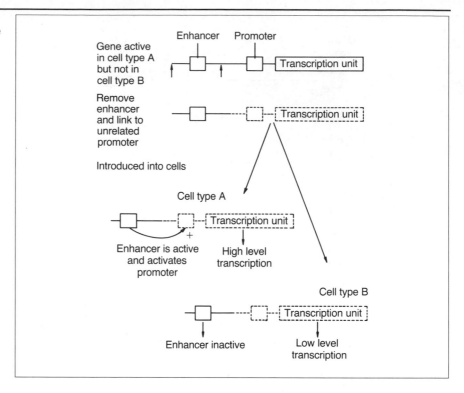

other tissues. The tissue-specific activity of these enhancer elements is likely to play a crucial role in the observed tissue-specific pattern of expression of the corresponding gene. Thus in the case of the insulin gene, early experiments involving linkage of different upstream regions of the gene to a marker gene, and subsequent introduction into different cell types, identified a region approximately 250 bases upstream of the transcriptional start site as being of crucial importance in producing high-level expression in pancreatic endocrine cells (Walker *et al.*, 1983). This position corresponds exactly to the position of the tissue-specific enhancer (Edlund *et al.*, 1985), indicating the importance of this element in gene regulation. Similarly, mutation of conserved sequences within the tissue-specific enhancers of genes expressed in the exocrine cells of the pancreas, such as elastase and chymotrypsin, abolishes the tissue-specific pattern of expression of these genes (Boulet *et al.*, 1986).

In the case of the insulin gene, the importance of the enhancer element in producing tissue-specific gene expression was further demonstrated by experiments in which this enhancer (together with its adjacent promoter) was linked to the gene encoding the large T antigen of the eukaryotic virus SV40, whose production can be measured readily using a specific antibody. The resulting construction was introduced into a fertilized mouse

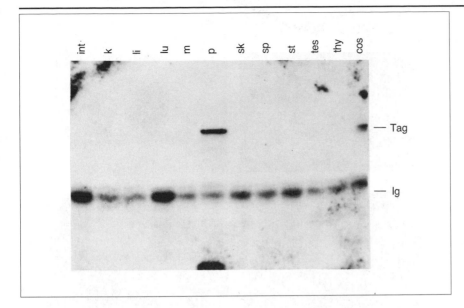

int | k | li | lu | m | p | sk | sp | st | tes | thy | cos

— Tag

— Ig

Figure 6.12 Assay for expression of a hybrid gene in which the SV40 T-antigen protein-coding sequence is linked to the insulin gene enhancer and promoter. The gene was introduced into a fertilized egg and a transgenic mouse containing the gene in every cell of its body isolated. Expression of the T antigen is assayed by immuno-precipitation of protein from each tissue with an antibody specific for T antigen. Note that expression of the T antigen is detectable only in the pancreas (p) and not in other tissues, indicating the tissue-specific activity of the insulin gene enhancer. The track labeled 'cos' contains protein isolated from a control cell line expressing T antigen. The Ig band in all tracks is derived from the immunoglobulin antibody used to precipitate the T antigen.

egg and the expression of large T antigen analyzed in all tissues of the transgenic mouse that developed following the return of the egg to the oviduct. Expression of large T was detectable only in the pancreas and not in any other tissue (Figure 6.12) and was observed specifically in the β cells of the pancreatic islets which produce insulin (Figure 6.13; Hanahan, 1985). The enhancer of the insulin gene is therefore capable of conferring the specific pattern of insulin gene expression on an unrelated gene *in vivo*.

Hence, enhancer elements constitute another type of DNA sequence which is involved in the activation of genes in a particular tissue or in response to a particular stimulus, as envisaged by the Britten and David-son model. In agreement with this idea, the binding of cell-type-specific proteins to the enhancer element has been demonstrated for many different enhancers, including those in the immunoglobulin (Sen and Baltimore, 1986) and insulin (Ohlsson and Edlund, 1986) genes.

Therefore, in many cases the tissue-specific expression of a gene will be determined both by the enhancer element and sequences adjacent to the promoter. In the liver-specific pre-albumin gene, for example, gene activity is controlled both by the promoter itself, which is active only in liver cells, and by an upstream enhancer element, which activates any promoter approximately 10-fold in liver cells and not at all in other cell types (Costa *et al.*, 1986). Similarly, in the immunoglobulin genes, when the enhancer and the promoter itself are separated, both exhibit B-cell-specific activity in isolation but the maximal expression of the gene is observed only when the two elements are brought together (Garcia *et al.*, 1986).

Figure 6.13 Immuno-fluorescence assay of pancreas preparations from the transgenic mice described in the legend to Figure 6.12 with antibodies to the indicated proteins. Note that the distribution of T antigen parallels that of insulin and not that of the other pancreatic proteins.

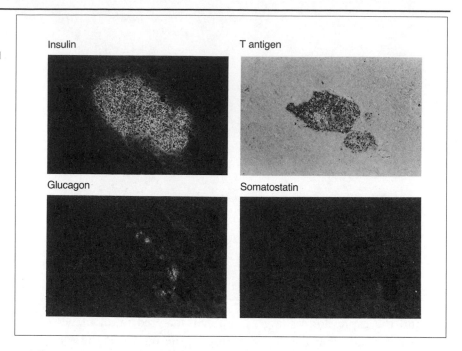

The importance of enhancer elements in the regulation of gene expression therefore necessitates consideration of the mechanism by which they act.

6.3.3   Mechanism of action of enhancers

In considering the nature of enhancers, we have drawn a distinction between these elements, which act at a distance, and the sequences discussed in Section 6.2, which are located immediately adjacent to the start site of transcription. In fact, however, closer inspection of the sequences within enhancers indicates that they are often composed of the same sequences found adjacent to promoters. For example, the immuno-globulin heavy-chain enhancers contain the octamer motif (ATGCAAAT) which, as discussed previously, is also found in the immunoglobulin promoters. The promoter and enhancer elements bind the identical B-cell-specific transcription factor (as well as a related protein found in all cell types) and play an important role in the B-cell-specific expression of the gene. Interestingly, within the enhancer the octamer motif is found within a modular structure containing binding sites for several different transcription factors, which act together to activate gene expression (Sen and Baltimore, 1986; Figure 6.14). In turn these modules can be subdivided into short DNA sequences, termed enhansons, which are the fundamental units of enhancer function (for review, see Dynan, 1989).

The close relationship of enhancer and promoter elements is further

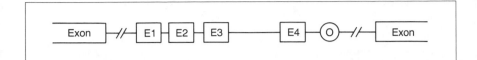

Figure 6.14 Protein-binding sites in the immunoglobulin heavy-chain gene enhancer. O indicates the octamer motif discussed in the text.

illustrated by the *Xenopus hsp70* gene, in which multiple copies of the heat-shock consensus element are located at positions far upstream of the start site and function as a heat-inducible enhancer element when transferred to another gene (Bienz and Pelham, 1986).

Enhancers therefore appear to consist of sequence motifs, which are also present in similarly regulated promoters and may be present within the enhancer associated with other control elements or in multiple copies. Indeed, the heat-shock consensus motif of the *Drosophila hsp70* gene, which we have used as the basic example of a promoter motif, has been shown to function as an enhancer when multiple copies are placed at a position well upstream of the transcriptional start site (Bienz and Pelham, 1986).

It seems likely, therefore, that enhancers may activate gene expression by either or both of the mechanisms described previously for promoter elements, i.e. a change in chromatin structure leading to nucleosome displacement or by direct interaction with the proteins of the transcriptional apparatus. In the case of chromatin structure changes, it is readily apparent that such changes caused by a protein binding to the enhancer could be propagated over large distances in both directions, causing the observed distance, position and orientation independence of the enhancer. In agreement with this possibility, DNaseI-hypersensitive sites have been mapped within a number of enhancer elements, including the immunoglobulin enhancer, and the nucleosome-free gap in the DNA of the eukaryotic virus SV40 (Figure 5.26) is located at the position of the enhancer.

At first sight, models involving the binding of protein factors to the enhancer followed by direct interaction with proteins of the transcriptional apparatus are more difficult to reconcile with the action at a distance characteristic of enhancers. None the less, the binding to enhancers of very many proteins crucial for transcriptional activation (reviewed by Hatzopoulos *et al.*, 1988) suggests that enhancers can indeed function in this manner. Models to explain this postulate that the enhancer serves as a site of entry for a regulatory factor. The factor would then make contact with the promoter-bound transcriptional apparatus either by sliding along the DNA, or via a continuous scaffold of other proteins, or by the looping out of the intervening DNA (Figure 6.15).

Of these possibilities, both the sliding model and the continuous

Figure 6.15 Possible models for the action of enhancers located at a distance from the activated promoter.

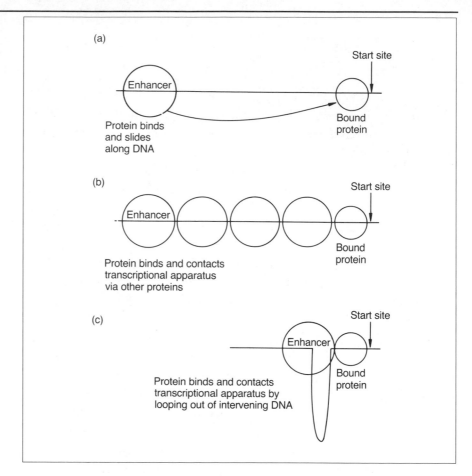

(a)

Start site

Enhancer

Protein binds
and slides
along DNA

Bound
protein

(b)

Start site

Enhancer

Protein binds and contacts
transcriptional apparatus
via other proteins

Bound
protein

(c)

Start site

Enhancer

Protein binds and contacts
transcriptional apparatus by
looping out of intervening DNA

Bound
protein

scaffold model are difficult to reconcile with the observed large distances over which enhancers act. Similarly, these models cannot explain the observations of Atchison and Perry (1986) who found that the immunoglobulin enhancer activates equally two promoters placed 1.7 and 7.7 kb away on the same DNA molecule, since they would postulate that sliding or scaffolded molecules would stop at the first promoter (Figure 6.16). Moreover, it has been shown that an enhancer can act on a promoter when the two are located on two separate DNA molecules linked only by a protein bridge, which would disrupt sliding or scaffolded molecules (Muller *et al.*, 1989).

Such observations are explicable, however, via a model in which proteins bound at the promoter and enhancer proteins have an affinity for one another and make contact via looping out of the intervening DNA. Moreover, such a model can explain readily the critical importance of DNA structure on the action of enhancers. Thus it has been shown that removal of precise multiples of 10 bases (one helical turn) from the region

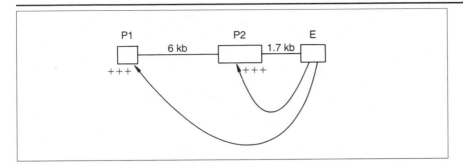

Figure 6.16 An enhancer
activates an adjacent
promoter and a more distant
one equally well.

between the SV40 enhancer and its promoter has no effect on its activity,
but deletion of DNA corresponding to half a helical turn disrupts
enhancer function severely (Takahasi *et al.*, 1986).

6.3.4   Positive and negative action of enhancer elements

Thus far we have assumed that enhancers act in an entirely positive
manner. In a tissue containing an active enhancer-binding protein,
the enhancer will activate a promoter, whereas in other tissues where the
protein is absent or inactive, the enhancer will have no effect. Such a
mechanism does indeed appear to operate for the majority of enhancers
which, when linked to promoters, activate gene expression in one or a few
cell types and have no effect in other cell types. In contrast, however, some
sequences appear to act in an entirely negative manner, inhibiting the
expression of genes which contain them. Following the initial identification of such an element in the cellular oncogene c-*myc*, similar elements,
which are referred to as silencers, have been defined in genes encoding
proteins as diverse as collagen type II, growth hormone and glutathione
transferase P (for review, see Jackson, 1991). Like enhancers, silencers can
act on distant promoters when present in either orientation but have an
inhibitory rather than a stimulatory effect on the level of gene expression.

As with enhancer elements it is likely that these silencers act either at the
level of chromatin structure by directing the tight packing of adjacent
DNA or by binding a protein which then directly inhibits transcription by
interacting with RNA polymerase and its associated factors. Examples of
silencers which appear to act in each of these ways have been observed
(Figure 6.17). Thus the silencer element located approximately 2 kb from
the promoters of the repressed mating type loci (Section 2.4) plays a
crucial role in organizing this region into the tightly packed structure
characteristic of non-transcribed DNA (for review, see Jackson, 1991).
Interestingly the silencer element appears to represent a site for attachment
of the DNA to the nuclear matrix. Hence, it may act by promoting the
further condensation of the chromatin solenoid to form a loop of DNA

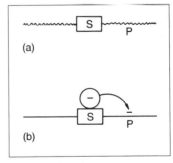

Figure 6.17 A silencer
element (S) can inhibit
activity of the promoter (P)
either (a) by organizing the
DNA into a tightly packed
chromatin structure or (b) by
binding an inhibitory
transcription factor (–).

Figure 6.18 The negative effect of the silencer element on transcription of the yeast mating type locus is eliminated when it is separated from the promoter by the DNA rearrangement event.

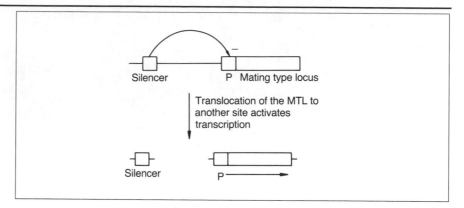

attached to the nuclear matrix which is the most condensed structure of chromatin (Section 5.3 and Figure 5.9).

In the case of the repressed mating type loci, relief from the negative effect of the silencer sequences is provided by the physical separation of the repressed locus from the silencer sequence and its translocation to another site, where it becomes active (Figure 6.18).

As well as acting at the level of chromatin structure, it is also possible for silencers to act by binding negatively acting transcription factors. Thus, the silencer in the gene encoding lysozyme appears to act, at least in part, by binding the thyroid hormone receptor which in the absence of thyroid hormone has a directly inhibitory effect on gene activity (Section 7.3.4).

In summary, therefore, whether acting positively or negatively, en-hancers play a critical role in the regulation of eukaryotic gene expression, often acting in concert with related regulatory elements located adjacent to the promoter itself.

6.4 Locus control regions

6.4.1 The locus control region

When specific genes are introduced into fertilized mouse eggs and used to create transgenic mice, the introduced gene is frequently expressed at very low levels and this expression is not increased when multiple copies of the genes are introduced. Similarly, different copies of the introduced gene which integrate into the host chromosomes at different positions are expressed at very different levels, suggesting that gene activity is being influenced by adjacent chromosomal regions. This effect occurs even when the gene is introduced together with its adjacent promoter and enhancer elements. This has led to the concept that some sequences,

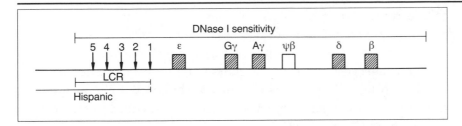

segment type="">**Figure 6.19** Organization of the β-globin gene locus showing the position of the locus control region (LCR) and the five DNaseI-hypersensitive sites within it (arrows). The region showing enhanced DNaseI sensitivity in the reticulocyte lineage is indicated together with the functional genes (hatched boxes) encoding epsilon globin (ε), the two forms of gamma globin (Gγ and Aγ), delta globin (δ) and beta globin (β) and the non-functional β-globin-like pseudogene (open box). The extent of the deletion in patients with Hispanic thalassemia is indicated by the line. Note that this deletion removes the LCR but leaves the genes themselves intact.

which are necessary for high-level gene expression independent of the position of the gene in the genome, are absent from the introduced gene.

This idea has been supported by studies in the β-globin gene cluster (Section 6.1.2) which contains four functional β-globin-like genes and a non-functional pseudogene (Figure 6.19). Thus a region located 10–20 kb upstream of the β-globin genes has been shown to confer high-level, position-independent expression when linked to a single globin gene and introduced into transgenic mice. Moreover, this element acts in a tissue-specific manner since its presence allows the globin gene to be expressed in the correct pattern, with high-level expression in erythroid cells and not in other cell types.

Most importantly this element is not only required for expression of the β-globin genes in transgenic animals but also plays a role in the natural expression of the globin genes. Thus its deletion in human individuals leads to a lack of expression of any of the genes in the cluster. This occurs even in cases where all the genes together with their promoter and enhancer regions remain intact, leading to a lethal disease known as Hispanic thalassemia in which no functional hemoglobin is produced (Figure 6.19).

The role of this element in stimulating activity of all the genes in the β-globin cluster has led to its being called a locus control region (LCR; for reviews, see Townes and Behringer, 1990; Dillon and Grosveld, 1993). As with enhancers, following its original definition in the β-globin cluster similar LCR elements have been identified in other genes, including the α-globin cluster, the major histocompatibility locus and the CD2 and lysozyme genes. It is clear therefore that LCRs constitute another important element essential for the correct regulation of gene expression.

6.4.2 Mechanism of action of LCRs

DNA sequence analysis of LCRs has indicated that they are rich in sequence motifs which are also found in promoter and enhancer elements and which could bind transcription factors. As with promoters and enhancers therefore it is in principle possible that the LCR could function either by direct interaction with the transcriptional apparatus or by affecting chromatin structure. In practice, however, the second of these is likely to be correct. Thus in many cases regions with LCR activity do not affect gene activity when introduced transiently into cells under conditions where the exogenous DNA is not packaged into chromatin but can do so when the same gene construct integrates into the host chromosome.

It is likely therefore that the LCR functions by affecting chromatin structure of the adjacent genes so that they can be transcribed (for review, see Felsenfeld, 1992). In agreement with this the region defined as the LCR on the basis of its functional activity contains five DNaseI-hypersensitive sites (Section 5.6) which appear early in erythroid development prior to the expression of any of the genes in the β-globin cluster (Figure 6.19). Moreover, the LCR is able to induce the adjacent region containing the β-globin genes to assume the enhanced DNaseI sensitivity which is characteristic of active or potentially active genes (Section 5.4) and the LCR marks the boundary of the region of the genome which shows this preferential sensitivity in erythroid cells (Figure 6.19). Thus the presence of this region would render the adjacent DNA capable of being expressed in a position-independent manner, allowing high-level, tissue-specific expression of the gene in transgenic mice regardless of the position in the genome into which it is integrated. In contrast, a gene lacking this sequence would lack this stimulatory action and would also be subject to the influence of adjacent regulatory elements which might inhibit its expression (Figure 6.20).

As with other DNaseI-hypersensitive sites (Section 5.6), the sites in the LCR are likely to be free of nucleosomes and may be in a highly super-coiled form subject to torsional stress. This may allow them to propagate a wave of altered chromatin structure to the adjacent region, displacing histone H1, thereby allowing the binding of HMG14 and 17 and causing the formation of the beads on a string structure characteristic of active or potentially active DNA throughout the adjacent region (Section 5.4). Interestingly, LCR sequences appear to be involved in the attachment of large domains or loops of chromatin to the nuclear matrix or scaffold (Section 5.3). Thus a region controlled by a single LCR such as that

Figure 6.20 An inserted gene containing an LCR will be organized into an open chromatin conformation characterized by enhanced DNaseI sensitivity compared to flanking DNA (a). In contrast an inserted gene lacking the LCR will be subject to repression by adjacent regions which direct its organization into a closed chromatin organization (b).

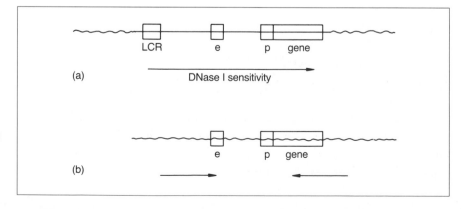

containing the β-globin genes may constitute a single large domain or loop which is attached to the nuclear scaffold and whose chromatin structure is regulated as a single unit.

Once an entire region of chromatin has been opened up in this way by the LCR, the promoter and enhancer elements of each individual gene in the region would direct its specific expression pattern. This is well illustrated in the β-globin gene cluster in which the DNaseI-hypersensitive sites within the LCR and the overall DNaseI sensitivity appear very early in erythroid development prior to the expression of any of the β-globin genes. Subsequently the ε-globin, γ-globin and β-globin genes are expressed successively in embryonic, fetal and adult erythroid cells, respectively, with expression being preceded in each case by the appearance of DNaseI-hypersensitive sites in the promoter/enhancer regions of the individual gene (Figure 6.21; for review, see Townes and Behringer, 1990). The LCR thus constitutes a control element which directs the chromatin structure of a large region of DNA, allowing the subsequent activation of individual genes within the cluster via the formation of DNaseI-hypersensitive sites and transcription factor binding at their individual promoter and enhancer regions (Figure 6.21).

6.5 Regulation of transcription by RNA polymerase I and III

As discussed in Section 6.1.2, two other polymerases exist each of which has its own specialized role in the cell. Interestingly, however, the TBP

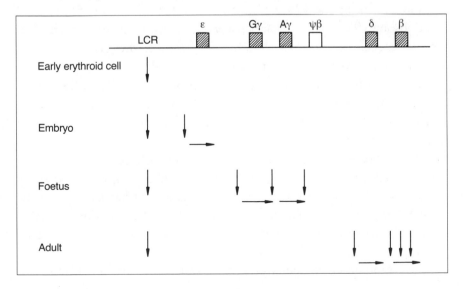

Figure 6.21 The appearance of multiple DNaseI-hypersensitive sites within the LCR (shown as a single arrow for simplicity) in early erythroid cells precedes the appearance of other hypersensitive sites adjacent to each individual gene which occurs later in development as these genes are sequentially expressed.

factor described in Section 6.2.2 is required for transcription by RNA polymerase I (Sharp, 1992) and RNA polymerase III (Rigby, 1993) as well as by RNA polymerase II. This has led to the suggestion that it may represent a fundamental factor which is essential for the basic process of transcription itself by interacting with all forms of RNA polymerase (White and Jackson, 1992). The specific activity of each polymerase on different sets of genes would then be dependent on the other factors in the transcriptional complex which are unique to each polymerase. The regulation of transcription by RNA polymerases I and II will now be discussed in turn.

6.5.1 RNA polymerase I

RNA polymerase I is responsible for the transcription of the tandem arrays of genes encoding ribosomal RNA, such transcription constituting about one-half of total cellular transcription. As with RNA polymerase II, regulatory sequences upstream of the start site of transcription are involved in controlling the transcription of these genes, with sequences from the initiation site to −50 playing an essential role and sequences farther upstream having a modulatory function (for reviews, see Sommerville, 1984; La Thangue and Rigby, 1988). Although the rate of transcription of the ribosomal genes can be altered by various stimuli, such as viral infection or changes in cellular growth rate, the processes whereby this occurs are not yet well understood.

6.5.2 RNA polymerase III

Of the many types of transcription unit that are transcribed by polymerase III, most attention has focused on the genes that encode the 5S RNA of the ribosome (for review, see Cilberto *et al.*, 1983). In an attempt to identify the sequences important for the expression of this gene, sequences surrounding it were deleted and the effect on the transcription of the gene in a cell-free system investigated (Sakonju *et al.*, 1980). Somewhat surprisingly, in view of the results with RNA polymerase I and II discussed above, the entire upstream region of the gene could be deleted with no effect on gene expression (Figure 6.22a and b). Indeed, deletions within the transcribed region of the 5S gene also had no effect on its expression until a boundary, 40 bases within the transcribed region, was crossed. By this means, an internal control region essential for the transcription of the 5S RNA gene was defined which, in contrast to the situation for other RNA polymerases, is located entirely within the transcribed region (Figure 6.23).

This region of the 5S gene was shown subsequently to bind a transcrip-

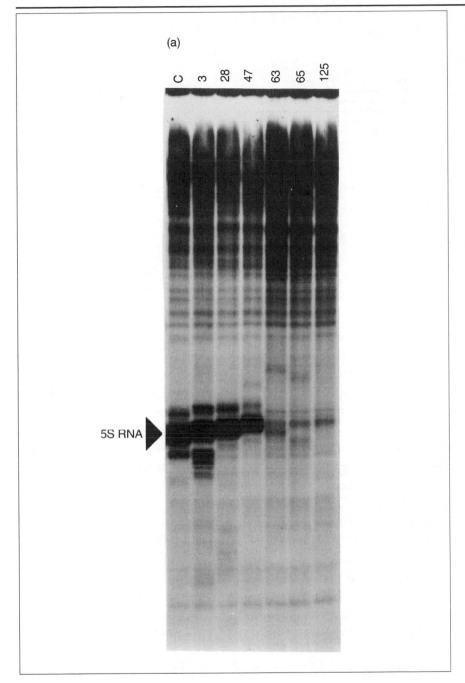

(a)

5S RNA ▶

C 3 28 47 63 65 125

Figure 6.22 Effect of deletions in the 5S rRNA gene on its expression. (a) Transcription assay in which the production of 5S RNA (arrowed) by an intact control 5S gene (C) and various deleted 5S genes is assayed. The numbers indicate the end point of each deletion used, 47 indicating that the deletion extends from the upstream region to the 47th base within the transcribed region, etc.

tion factor known as TFIIIA (for review, see Pieler and Theunissen, 1993) (Figure 6.24). The binding of this factor, which is needed only for transcription of the 5S gene and not for that of other RNA polymerase III genes, is the first step in transcription of this gene. Subsequently, the

Figure 6.22 (b) Summary of the extent of the deletions used and their effects on transcription. The use of these deletions allows the identification of a critical control element (boxed) within the transcribed region of the 5S gene.

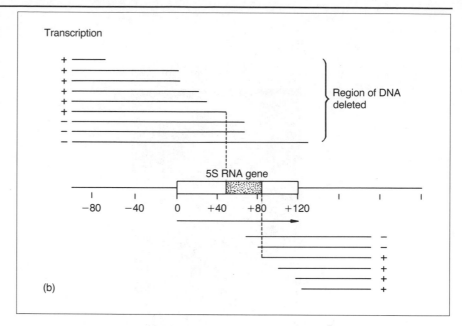

(b)

transcription factors TFIIIB and TFIIIC associate with TFIIIA and a stable transcription complex forms (Figure 6.25). This complex, which is stable through many cell divisions, promotes the binding of RNA polymerase III to the adjacent region of the gene (Figure 6.26) and transcription occurs. The binding of RNA polymerase III is dependent on the presence of the stable transcription complex and not on the precise sequence of the DNA to which it binds since, as discussed above, the region to which the polymerase normally binds can be deleted and replaced by other sequences without drastically reducing transcription.

Interestingly, despite the existence of a distinct type of promoter with an internal control region, the TBP factor is essential for the transcription of the 5S gene and other genes transcribed by RNA polymerase III, forming a part of the TFIIIB factor, which is a multi-protein complex consisting of TBP and other associated factors (Rigby, 1993).

The existence of internal control regions essential for transcription is not unique to the 5S rRNA gene. Internal promoter elements essential for

Figure 6.23 The control element of the 5S gene is contained within the transcribed region, resulting in its presence in the 5S RNA as well as the DNA.

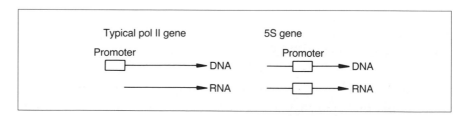

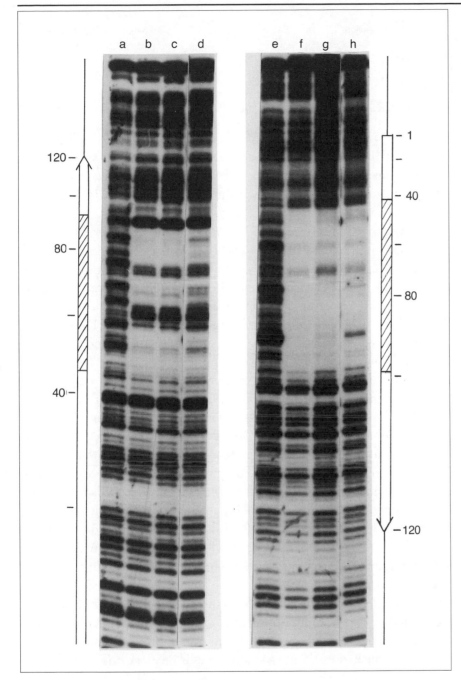

Figure 6.24 Binding of the TFIIIA transcription factor to the internal control region of 5S DNA in a footprint assay, in which binding of a protein protects the DNA from digestion by DNaseI and produces a clear region lacking the ladder of bands produced by DNaseI digestion of the other regions. Tracks a and e show the two DNA strands of the 5S gene in the absence of added TFIIIA, while tracks b–d and f–h show the same DNA in the presence of TFIIIA.

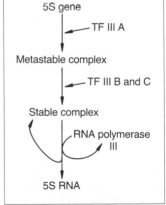

Figure 6.25 Stages in the formation of a stable transcription complex on the 5S gene.

transcription have also been identified in other genes transcribed by RNA polymerase III, such as the genes encoding the transfer RNAs, that encoding the 7SL RNA of the signal recognition particle, and the Alu repeated sequences. These internal sequences are recognized by the transcription

Figure 6.26 Binding of TFIIIA and of RNA polymerase III to the 5S gene.

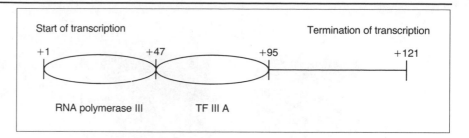

factors TFIIIB and TFIIIC discussed above. In some genes such internal sequences operate in conjunction with other sequences upstream of the transcribed region which may play the predominant role in individual cases (reviewed by Sollner-Webb, 1988).

Hence, genes transcribed by RNA polymerase III differ from those transcribed by RNA polymerases I and II in having sequences within the transcribed region which are involved in the initiation of transcription. In the 5S rRNA system, the fact that, unlike upstream sequences, such internal elements will be present in both the template DNA and the transcribed RNA (Figure 6.25) is exploited in a unique regulatory mechanism. The toad, *Xenopus laevis*, contains two types of 5S genes: the oocyte genes, which are transcribed only in the developing oocyte before fertilization, and the somatic genes, which are transcribed in cells of the embryo and adult. The internal control region of both these types of genes bind TFIIIA, whose binding is necessary for their transcription. However, sequence differences between the two types of gene (Figure 6.27) result in a higher affinity of TFIIIA factor for the somatic compared with the oocyte genes. Hence, the oocyte genes are only transcribed in the oocyte, where there are abundant levels of TFIIIA, and not in other cells, where the levels are only sufficient for activity of the somatic genes. In the developing oocyte TFIIIA is synthesized at high levels and transcription of the oocyte genes begins. As more and more 5S RNA molecules containing the TFIIIA-binding site accumulate in the maturing oocyte, they bind the transcription factor. This factor is thus sequestered with the 5S RNA into storage particles and is unavailable for transcription of the 5S rRNA genes. Hence the level of free transcription factor falls below that necessary for the transcription of the oocyte genes and their transcription ceases, while transcription of the higher affinity somatic genes is unaffected (Figure 6.28).

Figure 6.27 Sequence of the internal control element of the *Xenopus* somatic 5S genes. The two base changes in the oocyte 5S genes which result in a lower affinity for TFIIIA are indicated below the somatic sequence.

```
+46                                                             +98
TCGGAAGCCAAGCAGGGTCGGGCCTGGTTAGTACTTGGATGGGAGACCGCCTGG
       G  T
```

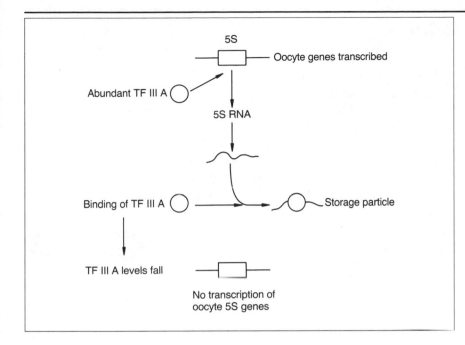

Figure 6.28 Sequestration of the TFIIIA factor by binding to 5S RNA made in the oocyte results in a fall in the level of the factor, switching off transcription of the oocyte-specific 5S genes. The somatic 5S genes, which have a higher affinity binding site for TFIIIA, continue to be transcribed.

A simple mechanism involving differential binding of a transcription factor to two related genes is thus used in conjunction with the binding of the same factor to its RNA product to modulate expression of the 5S genes.

6.6 Conclusions

In this chapter we have discussed a number of sequence elements, which can confer a particular pattern of gene regulation causing increased expression in response to a particular signal or in a particular tissue. Such sequences are present in a variety of locations close to or distant from the start site of transcription, in intervening sequences or, in the case of polymerase III transcription units, within the region encoding the RNA product itself. In all cases, however, such regions act by binding a protein which recognizes specific sequences within the control region. The activity of these proteins is crucial to the increased transcription mediated by the sequences to which they bind. The nature and function of these proteins is discussed in the next chapter.

References

Ashburner, M. and J. J. Bonner 1979. The induction of gene activity in *Drosophila* by heat shock. *Cell* 17, 241–54.

Atchison, M. L. and R. P. Perry 1986. Tandem kappa immunoglobulin promoters are equally active in the presence of the kappa enhancer: implications for models of enhancer function. *Cell* 46, 253–62.

Beato, M. 1989. Gene regulation by steroid hormones. *Cell* 56, 335–44.

Bienz, M. and H. R. B. Pelham 1986. Heat shock regulatory elements function as an inducible enhancer when linked to a heterologous promoter. *Cell* 45, 753–60.

Boulet, A. M., C. R. Erwin and W. J. Ruther 1986. Cell-specific enhancers in the rat exocrine pancreas. *Proceedings of the National Academy of Sciences of the USA* 83, 3599–603.

Breathnach, R. and P. Chambon 1981. Organization and expression of eucaryotic split genes coding for proteins. *Annual Review of Biochemistry* 52, 441–66.

Britten, R. J. and E. H. Davidson 1969. Gene regulation for higher cells: a theory. *Science* 165, 349–58.

Buratowski, S. 1994. The basis of transcription by RNA polymerase II. *Cell* 77, 1–3.

Cilberto, G., L. Castagnoli and R. Cortese 1983. Transcription by RNA polymerase III. *Current Topics in Developmental Biology* 18, 59–88.

Costa, R. H., E. Lai and J. E. Darnell, Jr 1986. Transcriptional control of the mouse prealbumin (transthyretin) gene: both promoter sequences and a distinct enhancer are cell specific. *Molecular and Cellular Biology* 6, 4697–708.

Davidson, E. H., H. T. Jacobs and R. J. Britten 1983. Very short repeats and coordinate induction of genes. *Nature* 301, 468–70.

Dillon, N. and F. Grosveld 1993. Transcriptional regulation of multigene loci, multilevel control. *Trends in Genetics* 9, 134–7.

Dynan, W. S. 1989. Modularity in promoters and enhancers. *Cell* 58, 1–4.

Edlund, T., M. D. Walker, J. Barr and W. J. Rutter 1985. Cell-specific expression of the rat insulin gene: evidence for role of two distinct 5' flanking elements. *Science* 230, 912–16.

Evans, J., G. Felsenfeld and M. Reitman 1990. Control of globin gene transcription. *Annual Review of Cell Biology* 6, 95–124.

Felsenfeld, G. 1992. Chromatin as an essential part of the transcriptional mechanism. *Nature* 355, 219–224.

Garcia, J. V., L. Bich-Thuy, J. Stafford and C. Queen 1986. Synergism between immunoglobulin enhancers and promoters. *Nature* 322, 383–5.

Gillis, S. D., S. L. Morrison, V. T. Oi and S. Tonegawa 1983. A tissue-specific transcription enhancer element is located in the major intron of a rearranged immunoglobulin heavy chain gene. *Cell* 33, 717–28.

Grosschedl, R. and M. L. Birnsteil 1980. Spacer DNA sequences up-stream of the TATAAATA sequence are essential for promotion of histone H2A transcription *in vivo*. *Proceedings of the National Academy of Sciences of the USA* 77, 7102–6.

Hanahan, D. 1985. Heritable formation of pancreatic beta-cell tumours in transgenic mice expressing recombinant insulin/simian virus 40 oncogenes. *Nature* 315, 115–22.

Hatzopoulos, A. K., U. Schlokat and P. Gruss 1988. Enhancers and other cis-acting sequences. In *Transcription and Splicing*, B. D. Hames and D. M. Glover (eds), 43–96. IRL Press, Oxford.

Hernandez, N. 1993. TBP: a universal transcription factor. *Genes and Development* 7, 1291–308.

Jackson, M.E. 1991. Negative regulation of eukaryotic transcription. *Journal of Cell Science* 1–7.

Jelinek, W. R. and C. W. Schmid 1982. Repetitive sequences in eukaryotic DNA and their expression. *Annual Review of Biochemistry* 55, 631–61.

Jones, N. C., P. W. J. Rigby and E. B. Ziff 1988. *Trans*-acting protein factors and the regulation of eukaryotic transcription. *Genes and Development* 2, 267–81.

Latchman, D. S. 1991. *Eukaryotic Transcription Factors*. Academic Press, London.

La Thangue, N. B. and P. W. J. Rigby 1988. *Trans*-acting protein factors and the regulation of eukaryotic transcription. In *Transcription and Splicing*, B. D. Hames and D. Glover (eds), 3–42. IRL Press, Oxford.

Lichtsteiner, S., J. Wuarin and U. Schibler 1987. The interplay of DNA binding proteins on the promoter of the mouse albumin gene. *Cell* 51, 9963–73.

McKnight, S. and R. Tjian 1986. Transcriptional selectivity of viral genes in mammalian cells. *Cell* 46, 795–805.

Maniatis, T., S. Goodbourn and J. A. Fischer 1987. Regulation of inducible and tissue-specific gene expression. *Science* 236, 1237–45.

Morimoto, R. 1993. Cells in stress: transcriptional activation of heat shock genes. *Science* 259, 1409–10.

Muller, M. M., T. Gerster and W. Schaffner 1988. Enhancer sequences and the regulation of gene transcription. *European Journal of Biochemistry* 176, 485–95.

Muller, H.-P., J. W. Soga and W. Schaffner 1989. An enhancer stimulates transcription in *trans* when linked to the promoter via a protein bridge. *Cell* 58, 767–77.

Ohlsson, H. and T. Edlund 1986. Sequence specific interactions of nuclear factors with the insulin gene enhancer. *Cell* 45, 35–44.

Parker, C. S. and J. Topol 1984. A *Drosophila* RNA polymerase II transcription factor binds to the regulatory site of an *hsp70* gene. *Cell* 37, 273–83.

Pelham, H. R. B. 1982. A regulatory upstream promoter element in the *Drosophila hsp70* heat-shock gene. *Cell* 30, 517–28.

Pieler, T. and O. Theunissen 1993. TFIIIA: nine fingers – three hands. *Trends in Biochemical Sciences* 18, 226–30.

Rigby, P. W. J. 1993. Three in one and one in three: it all depends on TBP. *Cell* 72, 7–10.

Roeder, R. G. 1991. The complexities of eukaryotic transcription initiation. *Trends in Biochemical Sciences* **16**, 402–8.

Sakonju, S., D. F. Bogenhagen and D. D. Brown 1980. A control region in the center of the 5S gene directs specific initiation of transcription. II. The 3' border of the region. *Cell* **19**, 27–35.

Schmitz, A. and D. Galas 1979. The interaction of RNA polymerase and *lac* repressor with the *lac* control region. *Nucleic Acids Research* **6**, 111–37.

Sen, R. and D. Baltimore 1986. Multiple nuclear factors interact with the immunoglobulin enhancer sequences. *Cell* **46**, 705–16.

Sentenac, A. 1985. Eukaryotic RNA polymerases. *CRC Critical Reviews in Biochemistry* **1**, 31–90.

Serfling, E., M. Jasin and W. Schaffner 1985. Enhancers and eukaryotic gene transcription. *Trends in Genetics* **1**, 224–30.

Sharp, P. 1992. TATA-binding protein is a classless factor. *Cell* **68**, 819–21.

Sollner-Webb, B. 1988. Surprises in polymerase III transcription. *Cell* **52**, 153–54.

Sommerville, J. 1984. RNA polymerase I promoters and cellular transcription factors. *Nature* **310**, 189–90.

Takahashi K., M. Vigneron, H. Matthes, A. Wildeman, M. Zeake and P. Chambon 1986. Requirement of stereospecific alignments for initiation from the simian virus 40 early promoter. *Nature* **319**, 121–6.

Thompson, C. C. and McKnight, S. L. 1992. Anatomy of an enhancer. *Trends in Genetics* **8**, 232–6.

Townes, T. M. and R. R. Behringer 1990. Human globin locus activation region (LAR): role in temporal control. *Trends in Genetics* **6**, 219–23.

van Holde, K. 1994. Properly preparing promoters. *Nature* **367**, 512–13.

Walker, M. D., T. Edlund, A. M. Boulet and W. J. Rutter 1983. Cell specific expression controlled by the 5' flanking region of the insulin and chymotrypsin genes. *Nature* **306**, 557–61.

Weis, L. and D. Reinberg 1992. Transcription by RNA polymerase II initiator directed formation of transcription-competent complexes. *FASEB Journal* **6**, 3300–9.

White, R. J. and S. P. Jackson 1992. The TATA-binding protein: a central role in transcription by RNA polymerases I, II and III. *Trends in Genetics* **8**, 284–8.

Williams, G. T., T. K. McClanahan and R. I. Morimoto 1989. Ela transactivation of the human *hsp70* promoter is mediated through the basal transcriptional complex. *Molecular and Cellular Biology* **9**, 2574–87.

Wirth, T., L. Staudt and D. Baltimore 1987. An octamer oligonucleotide upstream of a TATA motif is sufficient for lymphoid specific promoter activity. *Nature* **329**, 174–8.

Zawel, L. and D. Reinberg 1992. Advances in RNA polymerase II transcription. *Current Opinion in Cell Biology* **4**, 488–95.

Zimarino, V. and C. Wu 1987. Induction of sequence-specific binding of *Drosophila* heat-shock activator protein without protein synthesis. *Nature* **327**, 727–30.

Transcriptional control – transcription factors

7

7.1 Introduction

As discussed in Chapter 6, the expression of specific genes in particular cell types or tissues is regulated by DNA sequence motifs present within promoter or enhancer elements which control the alteration in chromatin structure of the gene that occurs in a particular lineage or the subsequent induction of gene transcription. It has been assumed for many years that such sequences would act by binding a regulatory protein which was only synthesized in a particular tissue or was present in an active form only in that tissue. In turn, the binding of this protein would result in the observed effect on gene expression.

The isolation and characterization of such factors proved difficult, however, principally because they were present in very small amounts. Hence, even if they could be purified, the amounts obtained were too small to provide much information as to the properties of the protein.

This obstacle was overcome by the cloning of the genes encoding a number of different transcription factors. Two general approaches were used to achieve this. In one approach (Figure 7.1), exemplified by the work of Kadonaga and Tjian (1986), the transcription factor Spl was purified by virtue of its ability to bind to its specific DNA binding site. The partial amino acid sequence of the protein was then obtained from the small amount of material isolated and was used in conjunction with the genetic code to predict a set of DNA oligonucleotides, one of which would encode this region of the protein. The oligonucleotides were then hybridized to a cDNA library prepared from Spl-containing HeLa cell mRNA. A cDNA clone derived from the Spl mRNA must contain the sequence capable of encoding the protein and hence will hybridize to the probe. In this experiment one single clone derived from the Spl mRNA was isolated by screening a library of 1 million

Figure 7.1 Isolation of cDNA clones for the Spl transcription factor by screening with short oligonucleotides predicted from the protein sequence of Spl. Because several different triplets of bases can code for any given amino acid, multiple oligonucleotides that contain every possible coding sequence are made. Positions at which these oligonucleotides differ from one another are indicated by the brackets containing more than one base.

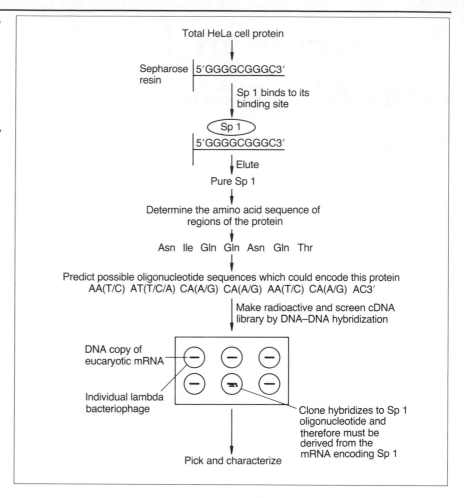

recombinants prepared from the whole population of HeLa cell mRNAs (Kadonaga *et al.*, 1987).

An alternative, more direct, approach to the cloning of transcription factors is exemplified by the work of Singh *et al.* (1988) on the NF κB protein, which is involved in regulating the expression of the immuno-globulin genes in B cells (Figure 7.2). As in the previous method, a cDNA library was constructed containing copies of all the mRNAs in a specific cell type. However, the library was constructed in such a way that the sequences within it would be translated into their corresponding proteins. This was achieved by inserting the cDNA into the coding region of the bacteriophage β-galactosidase gene, resulting in the translation of the eukaryotic insert as part of the bacteriophage protein (for a full description of this technique and its other applications, see Young and Davis, 1983). Most interestingly, these fusion proteins were capable of binding DNA with the same specificity as the original transcription factor encoded by

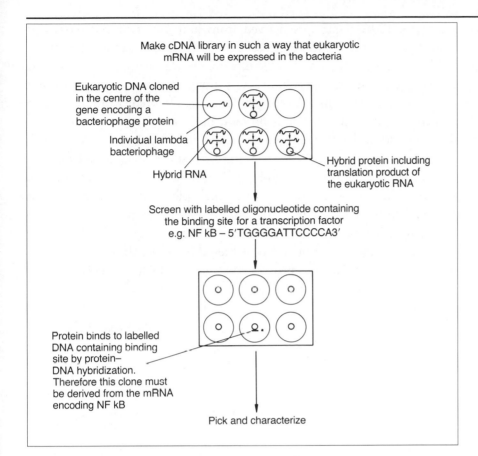

Figure 7.2 Isolation of cDNA clones for the NF κB transcription factor by screening an expression library with a DNA probe containing the binding site for the factor.

Make cDNA library in such a way that eukaryotic mRNA will be expressed in the bacteria

Eukaryotic DNA cloned in the centre of the gene encoding a bacteriophage protein

Individual lambda bacteriophage

Hybrid RNA

Hybrid protein including translation product of the eukaryotic RNA

Screen with labelled oligonucleotide containing the binding site for a transcription factor e.g. NF kB – 5′TGGGGATTCCCCA3′

Protein binds to labelled DNA containing binding site by protein– DNA hybridization. Therefore this clone must be derived from the mRNA encoding NF kB

Pick and characterize

the cloned mRNA. Hence the library could be screened directly with the radiolabeled DNA binding site for a particular transcription factor. A clone containing the mRNA for this factor, and hence expressing it as a fusion protein, bound the labeled DNA and could be identified readily and isolated.

Unlike the previous method, this procedure involves DNA–protein rather than DNA–DNA binding and can be used without prior purification of the transcription factor, provided its binding site is known. Since most factors are identified on the basis of their binding to a particular site, this is not a significant problem and the use of these two methods has resulted in the isolation of the genes encoding a wide variety of transcription factors.

In turn, this has resulted in an explosion of information on these factors (for general reviews, see Johnson and McKnight, 1989; Mitchell and Tjian, 1989; Latchman, 1991). Thus, once the gene for a factor has been cloned, Southern blotting (Section 2.2.4) can be carried out to study the structure of the gene, Northern blotting (Section 1.3.2) can be used to search for RNA transcripts derived from it in different cell types and related genes expressed in other tissues or other species can be identified.

More importantly, considerable information can be obtained from the cloned gene about the corresponding protein and its activity. Thus, not only can the DNA sequence of the gene be used to predict the amino acid sequence of the corresponding protein, but the existence of functional domains within the protein with particular activities can also be defined. As described above, if the gene encoding a transcription factor is expressed in bacteria, it continues to bind DNA in a sequence-specific manner. Hence if the gene is broken up into small pieces and each of these is expressed in bacteria (Figure 7.3), the abilities of each portion to bind to DNA, to other proteins or to a potential regulatory molecule can be assessed. This mapping can also be achieved by transcribing and translating pieces of the DNA into protein fragments in the test tube and testing their activity in the same way.

Figure 7.3 Mapping of the DNA-binding region of a transcription factor by testing the ability of different regions to bind to the appropriate DNA sequence when expressed in bacteria.

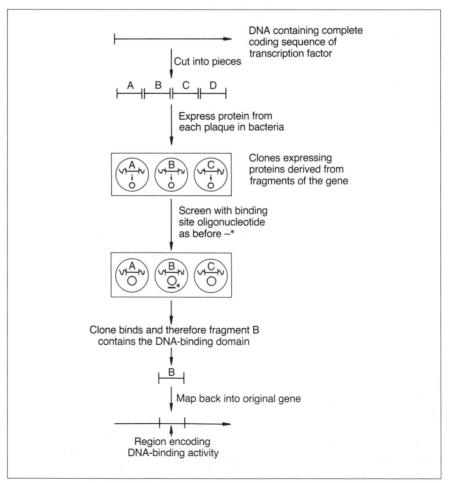

Each of the domains identified in this way can be altered by mutagenesis of the DNA and subsequent expression of the mutant protein as before. The testing of the effect of these mutations on the activity mediated by the particular domain of the protein will thus allow the identification of the amino acids that are critical for each of the observed properties of the protein.

In this way large amounts of information have accumulated on individual transcription factors. Rather than attempt to consider each factor individually, we will focus on the properties necessary for such a factor and illustrate our discussion by referring to the manner in which these are achieved in individual cases.

It should be evident from the foregoing discussion that the first property such a factor requires is the ability to bind to DNA in a sequence-specific manner and this is discussed in Section 7.2. Subsequently, the bound factor must influence transcription either positively or negatively by interacting with other transcription factors or with the RNA polymerase itself. Section 7.3 therefore considers the means by which DNA-bound transcription factors actually regulate transcription. Finally, in the case of transcription factors which activate a particular gene in one tissue only, some means must be found to ensure that the transcription factor is active only in that tissue. Section 7.4 discusses how this is achieved, either by the expression of the gene encoding the factor only in one particular tissue or by a tissue-specific modification which results in the activation of a factor present in all cell types.

7.2 DNA binding by transcription factors

7.2.1 Introduction

Extensive studies of eukaryotic transcription factors have identified several structural elements, which either bind directly to DNA or which facilitate DNA binding by adjacent regions of the protein (for reviews, see Harrison, 1991; Latchman, 1991; Travers, 1993). These motifs will be discussed in turn, using transcription factors that contain them to illustrate their properties.

7.2.2 The helix-turn-helix motif

The homeobox
The small size and rapid generation time of the fruit fly *Drosophila melanogaster* has led to it being one of the best-characterized organisms

Figure 7.4 Effect of a homeotic mutation, which produces a middle leg (b) in the region that would contain the antenna of a normal fly (a). al, all and alll, first, second and third antennal segments; ar, arista; ta, tarsus; ti, tibia; fe, femur; ap, apical bristle.

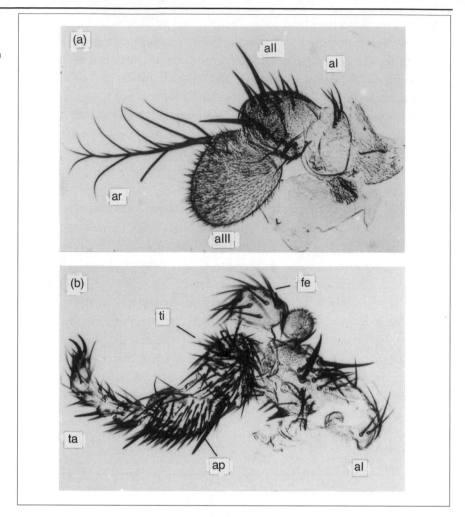

genetically and a number of mutations which affect various properties of the organism have been described. These include mutations which affect the development of the fly, resulting, for example, in the production of additional legs in the position of the antennae (Figure 7.4). Genes of this type are likely to play a crucial role in the development of the fly and, in particular, in determining the body plan, and are known as homeotic genes (for reviews, see Ingham, 1988; Scott and Carroll, 1987).

The critical role for the products of these genes, identified genetically, suggested that they would encode regulatory proteins which would act at particular times in development to activate or repress the activity of other genes encoding proteins required for the production of particular structures. This idea was confirmed when the genes encoding these proteins were cloned. Thus, these proteins were shown to be able to bind to

DNA in a sequence-specific manner and to be able to induce increased transcription of genes which contained this binding site (for review, see Hayashi and Scott, 1990). Thus in the case of the homeotic gene *fushi tarazu* (*ftz*), mutation of which produces a fly with only half the normal number of segments, the protein has been shown to bind specifically to the sequence TCAATTAAATGA. When the gene encoding this protein is introduced into *Drosophila* cells with a marker gene containing this sequence, transcription of the marker gene is increased. This up-regulation is entirely dependent on binding of the Ftz protein to this sequence in the promoter of the marker gene, since a 1 bp change in this sequence, which abolishes binding, also abolishes the induction of transcription (Figure 7.5).

The product of another homeotic gene, the engrailed protein, binds to the identical sequence to that bound by Ftz. Its binding does not produce increased transcription of the marker gene, however, and indeed it prevents the activation by Ftz. Hence, the expression of Ftz alone in a cell would activate particular genes, whereas Ftz expression in a cell also expressing the engrailed product would have no effect (Figure 7.6). In this way interacting homeotic gene products expressed in particular cells could control the developmental fate of the cells (for review, see Edelman and Jones, 1993).

Interestingly, there is evidence that the homeotic genes may be necessary not only for the actual production of a specific cell type but also for the long-term process of commitment to a particular cellular phenotype which was discussed in Chapter 5 (Section 5.2). Thus, in the case of the imaginal disks of *Drosophila* (Section 5.2), commitment to the production of a particular adult structure was maintained through many cell generations in the absence of differentiation. If during this time, however, a mutation is introduced into one of the homeotic genes in the disk cells, inactivating it, when eventually the cells are allowed to differentiate they will produce the wrong structure. Thus, for example, if the homeotic gene

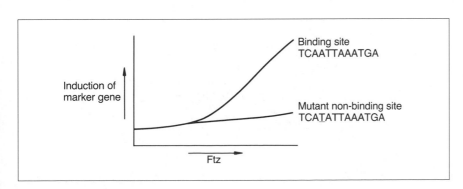

Figure 7.5 Effect of expression of the Ftz protein on the expression of a gene containing its binding site, or a mutated binding site containing a single base pair change which abolishes binding of Ftz.

Figure 7.6 Blockage of gene induction by Ftz in cells expressing the engrailed (Eng) protein which binds to the same sequence as Ftz but does not activate transcription.

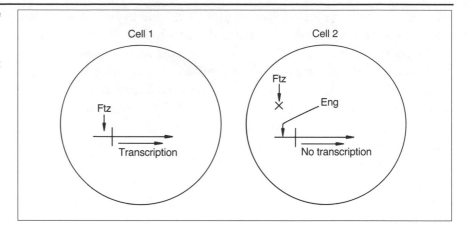

ultrabithorax (*ubx*) is inactivated in a disk cell which normally gives rise to the haltere (balancer), these cells will produce wing tissue when allowed to differentiate. Thus the continual expression of homeotic genes within the cells is essential for their commitment to a particular pathway of differentiation.

A possible molecular mechanism for this is provided by the demonstration (Krasnow *et al.*, 1989) that the Ubx protein binds to its own promoter and up-regulates its own transcription. Hence once production of this protein has been induced, presumably during the commitment process, it will continue indefinitely and thus maintain this commitment (Figure 7.7). Interestingly, however, as noted in Chapter 5 (Section 5.2), the changes in commitment which occur in imaginal disks when this process breaks down are precisely those which occur in homeotic mutations. Hence a change in the chromatin structure and expression of a specific homeotic gene in the imaginal disk will result in a change in the pattern of commitment similar to that which occurs when this gene is mutated.

The clear evidence that homeotic gene products regulate both their own genes and other genes by binding specifically to DNA has led to extensive investigation of their structure in order to identify the region that mediates

Figure 7.7 The Ubx protein activates its own promoter, producing a positive feedback loop maintaining high-level production of Ubx.

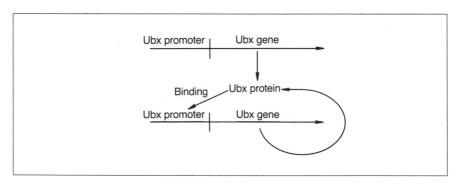

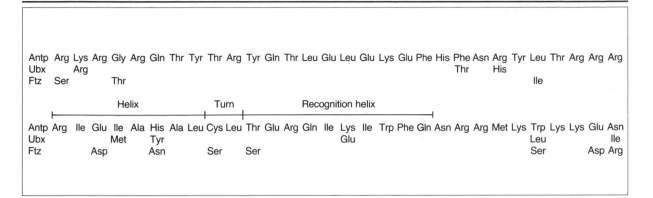

Figure 7.8 Amino acid sequences of several *Drosophila* homeodomains, showing the conserved helical motifs. Differences between the sequences of the Ubx and Ftz homeodomains from those of Antp are indicated; a blank denotes identity in the sequence. The helix-turn-helix region is indicated.

this sequence-specific DNA binding. When the genes encoding these proteins were first cloned, it was found that they each contained a short related DNA sequence of about 180 bp capable of encoding 60 amino acids (Figure 7.8), which was flanked on either side by sequences that differed dramatically between the different genes. The presence of this sequence, which was named the homeobox or homeodomain (for reviews, see Gehring, 1987 Scott *et al.*, 1989; Kornberg, 1993), in all these genes suggested that it plays a critical role in mediating their regulatory function. This suggestion was confirmed subsequently by the use of the homeobox as a probe to isolate other previously uncharacterized *Drosophila* regulatory genes.

The idea that the homeodomain might be of importance in the DNA binding and transcriptional activation function of the homeotic proteins was supported by the discovery of this element also in the products of the yeast mating type loci, the a and α proteins (Shepherd *et al.*, 1984). Thus, as discussed in Chapter 2 (Section 2.4), gene rearrangement results in the production of either the a or α protein in each individual yeast. These proteins then bind to DNA in a sequence-specific manner and transcriptionally activate other genes to produce the a or α mating type phenotype.

Hence the finding of the homeodomain in these proteins also strongly implicated it in the process of DNA binding. This hypothesis has been confirmed directly by the synthesis of the homeobox region of the Antennapedia protein without the remainder of the protein, either by expression in bacteria or by chemical synthesis and showing that it can bind to DNA in the identical sequence-specific manner to that exhibited by the intact protein.

The localization of this DNA binding to a short region of the protein only 60 amino acids in length allows a detailed structural prediction of the corresponding protein, which reveals that it contains a so-called helix-turn-helix motif which is highly conserved between the different

Figure 7.9 The helix-turn-helix motif.

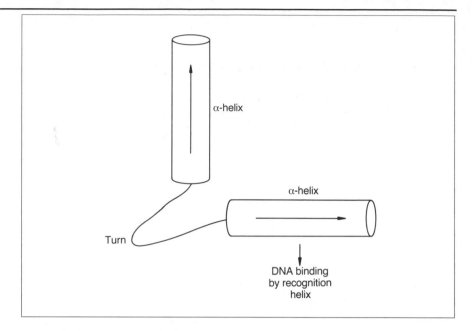

Drosophila proteins and the yeast mating type gene products (for review, see Affolter *et al.*, 1990). In this motif, a short region which can form an α-helical structure is followed by a β-turn and then another α-helical region. The position of these elements in the homeodomain is shown in Figure 7.8 and a diagram of the helix-turn-helix motif is given in Figure 7.9.

The prediction that this structure exists in the DNA-binding homeodomain has recently been directly confirmed by X-ray crystallographic analysis. Moreover, by carrying out this analysis on the homeodomain bound to its DNA binding site, it has been shown that the helix-turn-helix motif does indeed contact DNA, with the second helix lying partly within the major groove where it can make specific contacts with the bases of the DNA (Fig. 7.10; Kissinger *et al.*, 1990). This second helix (labeled the recognition helix in the homeobox sequence in Fig. 7.8) can thus mediate sequence-specific binding.

The presence of this structure therefore indicates how the homeobox proteins can bind specifically to particular DNA binding sequences, which is the first step in transcriptional activation of their target genes. The role of the helix-turn-helix motif in the recognition of specific sequences in the DNA has been demonstrated directly (Hanes and Brent, 1989). Thus a mutation which changes a lysine at position nine of the recognition helix in the Bicoid protein to the glutamine found in the equivalent position of the Antennapedia protein results in the protein binding to DNA with the sequence specificity of an Antennapedia rather

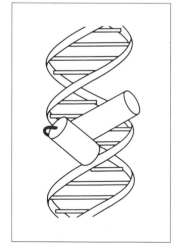

Figure 7.10 Binding of the helix-turn-helix motif to DNA, with the recognition helix in the major groove of the DNA.

than a Bicoid protein. Hence, not only does the helix-turn-helix motif mediate DNA binding, but differences in the precise sequence of this motif in different homeoboxes control the precise DNA sequence to which these proteins bind. Clearly, further structural and genetic studies of how this is achieved will throw considerable light on the way in which these proteins function.

The POU domain

The obvious importance of the homeobox and its presence in organisms as diverse as *Drosophila* and yeast prompted a search for proteins containing this element in other organisms. Both *Xenopus* and mammals have been shown to contain proteins of this type, which are expressed in specific cell types in the early embryo and are likely to play a regulatory role (for review, see Akam, 1989). Hence these proteins may provide a link between the processes regulating development in a variety of different organisms.

More recently, however, another class of regulatory proteins has been identified which contains the homeobox as one part of a much larger, 150–160 amino acid, conserved region known as the POU domain (for reviews, see Verrijzer and Van der Vliet, 1993; Wegner *et al.*, 1993). Unlike the homeobox proteins, these regulatory proteins were not identified by mutational analysis or by homology to other regulatory proteins, but were characterized as transcription factors having a particular pattern of activity.

Thus the mammalian Oct-1 and Oct-2 proteins both bind to the octamer sequence (consensus ATGCAAATNA) in the promoters of genes such as the histone H2B gene and those encoding the immunoglobulins, and mediate the transcriptional activation of these genes (for review, see Schaffner, 1989). When the genes encoding these proteins were cloned they were found to possess a 150–160 amino acid sequence that was also found in the mammalian Pit-1 protein, which regulates gene expression in the pituitary by binding to a sequence related to but distinct from the octamer (Ingraham *et al.*, 1988), and in the protein encoded by the nematode gene *unc*-86, which is involved in sensory neuron development.

This POU (Pit–Oct–Unc) domain contains both a homeobox-like sequence and a second conserved domain, the POU-specific domain (Figure 7.11). Although there are some differences between different POU proteins, in general the isolated homeodomain of the POU proteins alone is sufficient for sequence-specific DNA binding, but unlike the classical homeobox, the binding is of relatively low affinity in the absence of the POU-specific domain. Hence both parts of the POU domain are required for high-affinity sequence-specific DNA binding, indicating that the POU

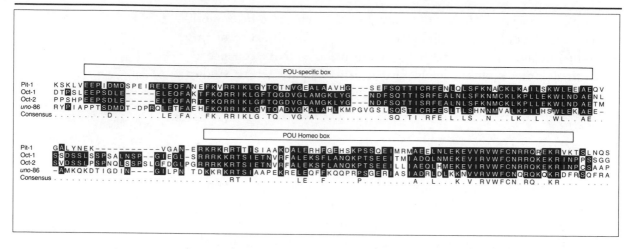

Figure 7.11 Amino acid sequences of the POU proteins. The homeodomain and the POU-specific domain are indicated. Solid boxes indicate regions of identity between the different POU proteins. The final line shows a consensus sequence obtained from the four proteins.

homeodomain and the POU-specific domain form two parts of a DNA-binding element which are held together by a flexible linker sequence.

Interestingly, like the POU homeodomain, the POU-specific domain can form a helix-turn-helix motif. It has therefore been suggested that the recognition helix from the POU-specific domain and that from the POU homeodomain bind to adjacent regions within the major groove of the DNA.

It is clear, therefore, that the POU proteins represent a new family of proteins, related to the homeobox proteins, which are likely to play a critical role in development. Thus inactivation of the Pit-1 gene leads to a failure of pituitary gland development resulting in dwarfism in mouse and humans whilst the *unc*-86 mutation results in a failure to form specific neurons in the nematode.

7.2.3 The zinc finger motif

The two cysteine–two histidine zinc finger
As discussed in Chapter 6 (Section 6.5.2), one of the earliest gene regulatory systems to be characterized was that of the gene encoding the 5S RNA of the ribosome, in which a transcription factor, TFIIIA, binds to the internal control region of the gene. This transcription factor was amongst the first to be purified (Miller *et al.*, 1985). The pure protein was shown to have a periodic repeated structure and to contain between seven and 11 atoms of zinc associated with each molecule of the pure protein.

The basis for this repeated structure was revealed when the gene encoding this protein was cloned and used to predict the corresponding amino

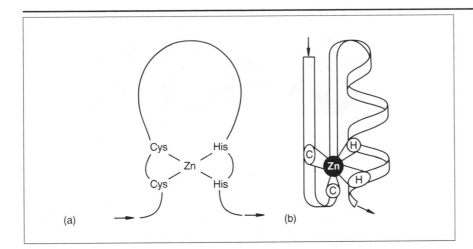

(a)   (b)

Figure 7.12 Two alternative possible structures of the two cysteine–two histidine zinc finger.

acid sequence. This protein sequence contained nine repeats of a 30 amino acid sequence of the form Tyr/Phe-X-Cys-X-Cys-$X_{2,4}$-Cys-X_3-Phe-X_5-Leu-X_2-His-$X_{3,4}$-His-X_5, where X is a variable amino acid. This repeating structure therefore contains two invariant pairs of cysteine and histidine residues which were predicted to bind a single zinc atom, accounting for the multiple zinc atoms bound by the purified protein.

This 30 amino acid repeating unit is referred to as a zinc finger, on the basis of the proposed structure in which a loop of 12 amino acids, containing the conserved leucine and phenylalanine residues as well as several basic residues, projects from the surface of the protein and is anchored at its base by the conserved cysteine and histidine residues, which directly co-ordinate an atom of zinc (Figure 7.12A). The binding of zinc by the cysteine and histidine residues has been confirmed directly by X-ray crystallographic analysis of the TFIIIA protein, although an alternative structure for the finger, involving an antiparallel β-sheet and α-helix between the zinc co-ordination sites, has also been proposed (Figure 7.12B).

Following the initial identification of the repetitive zinc fingers in TFIIIA, it was hypothesized (Miller *et al.*, 1985) that the tips of the fingers would directly contact the 5S DNA, this region being rich in basic amino acids such as arginine and lysine, which could potentially interact with the acidic DNA. The repetitive structure contains multiple fingers which allow the relatively small TFIIIA protein to make repeated contacts with the DNA along the relatively large (approximately 50 bp) 5S DNA internal control region. Subsequent experiments have confirmed this view and have suggested that each TFIIIA finger binds in the major groove of the DNA helix, interacting with five bases of DNA, or half a helical turn, with successive fingers binding on opposite sides of the

Figure 7.13 Model of the binding of the zinc fingers in TFIIIA to the 5S DNA. Note that adjacent fingers make contact with the DNA from opposite sides of the helix.

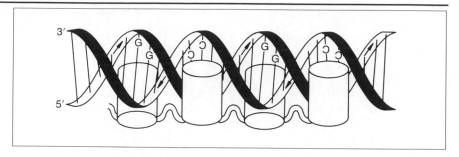

helix (Figure 7.13; reviewed by Klug and Rhodes, 1987; Rhodes and Klug, 1993).

It is clear, therefore, that like the helix-turn-helix motif, the zinc finger represents a protein structure capable of mediating the DNA binding of transcription factors. Although originally identified in the RNA polymerase III transcription factor TFIIIA, this motif has now been identified in a number of RNA polymerase II transcription factors and shown to play a critical role in their ability to bind to DNA and thereby influence transcription (Table 7.1; for reviews, see Evans and Hollenberg, 1988; Struhl, 1989).

Thus three contiguous copies of the 30 amino acid zinc finger motif are found in the transcription factor Spl whose cloning was discussed earlier (Section 7.1). The sequence-specific binding pattern of the intact Spl protein can be reproduced by expressing in *Escherichia coli* a truncated protein containing only the zinc finger region, confirming the importance of this region in DNA binding (Kadonaga *et al.*, 1987). Similarly, the *Drosophila*

Table 7.1 Transcriptional regulatory proteins containing Cys_2His_2 zinc fingers

Organism	Gene	Number of fingers
Drosophila	*kruppel*	4
	hunchback	6
	snail	4
	glass	5
Yeast	*ADR1*	2
	Swi5	3
Xenopus	*TFIIIA*	9
	Xfin	37
Mammal	*NGF-1a (Egr1)*	3
	MK1	7
	MK2	9
	Evi 1	10
	Sp1	3

Kruppel protein, which is vital for proper thoracic and abdominal development, contains four zinc finger motifs. A single mutation, which results in the replacement of the conserved cysteine in one of these fingers by a serine which could not bind zinc, leads to the complete abolition of the function of the protein, resulting in a mutant fly whose appearance is indistinguishable from that produced by complete deletion of the gene (Redemann *et al.*, 1988).

The zinc finger therefore represents a DNA-binding element which is present in variable numbers in many regulatory proteins. Indeed, the linkage between the presence of this motif and the ability to regulate gene expression is now so strong that, as with the homeobox, it has been used as a probe to isolate the genes encoding new regulatory proteins. The Kruppel zinc finger, for example, has been used in this way to isolate Xfin, a 37 finger protein expressed in the early *Xenopus* embryo (Ruiz-i-Altaba u*et al.*, 1987).

The possible involvement of zinc finger proteins in controlling development in vertebrates is mirrored in *Drosophila*, where numerous proteins involved in regulating development, such as Kruppel, Hunchback and Snail, contain zinc fingers. The interactions of these proteins with the homeobox proteins, which contain the alternative DNA-binding helix-turn-helix motif, is of central importance in the development of *Drosophila* and possibly other organisms.

The multi-cysteine zinc finger

Throughout this work we have noted that the effect of steroid hormones on mammalian gene expression is one of the best-characterized examples of gene regulation. Thus the steroid-regulated genes were amongst the first to be shown to be regulated at the level of gene transcription (Chapter 3) by means of the binding of a specific receptor to a specific DNA sequence (Section 6.2), resulting in the displacement of a nucleosome and the generation of a DNaseI-hypersensitive site (Section 5.6). When the genes encoding the DNA-binding receptors for the various steroid hormones, such as glucocorticoid and estrogen, were cloned, they were found to constitute a family of proteins encoded by distinct but related genes. In turn, these proteins were related to other receptors which mediated the response of the cell to hormones such as thyroid hormone or retinoic acid, leading to the idea of an evolutionarily related family of genes encoding hormone receptors, known as the steroid–thyroid hormone receptor gene superfamily (for reviews, see Evans, 1988; Beato, 1989; Parker, 1993).

When the detailed structures of the members of this family were compared (Figure 7.14) it was found that each had a multi-domain structure,

Figure 7.14 Domain structure of individual members of the steroid–thyroid hormone receptor super-family. The proteins are aligned on the DNA-binding domain, which shows the most conservation between different receptors. The percentage homologies in each domain of the receptors to that of the glucocorticoid receptor are indicated.

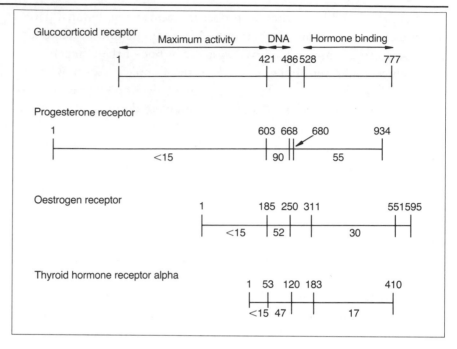

which included a central highly conserved domain. On the basis of experiments in which truncated versions of the receptors were introduced into cells and their activities measured, it was shown that this conserved domain mediated the DNA-binding ability of the receptor, while the C-terminal region was involved in the binding of the appropriate hormone and the N-terminal region was involved in producing maximal induction of transcription of target genes (for an example of this type of approach, see Hollenberg *et al.*, 1987).

Sequence analysis of the DNA-binding domain in a variety of receptors showed that it conformed to a consensus sequence of the type $Cys-X_2$-$Cys-X_{13}$-$Cys-X_2$-$Cys-X_{15,17}$-$Cys-X_5$-$Cys-X_9$-$Cys-X_2$-$Cys-X_4$-Cys. Like the cysteine–histidine finger described in the previous section, the DNA binding of this element is dependent upon the presence of zinc or a related heavy metal such as cadmium. Moreover, this element can be drawn as two conventional zinc fingers, in which four cysteines replace the two cysteine–two histidine structure of the conventional finger in binding zinc and which are separated by a linker region containing the 15–17 variable amino acids (Figure 7.15). Such a structure is supported by spectrographic analysis of this region of the receptor, which clearly demonstrates the presence of two zinc atoms each co-ordinated by four cysteines in a tetrahedral array. However, such structural analysis also indicates that the two fingers in the steroid receptors constitute a single

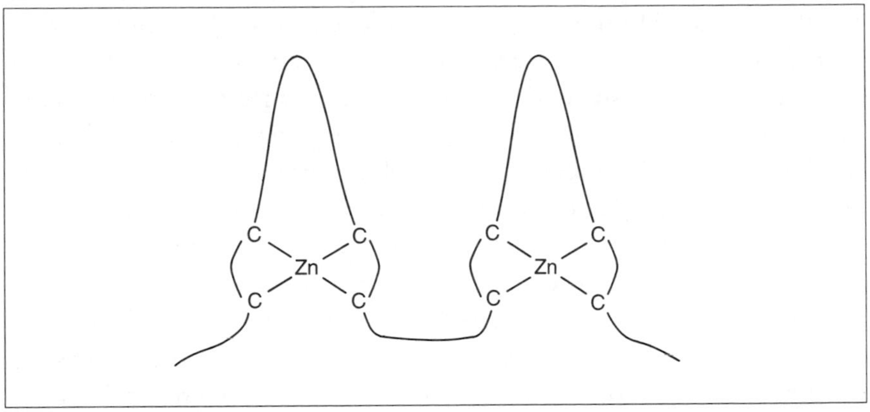

Figure 7.15 Structure of the four-cysteine zinc finger.

structural element, unlike the situation in the cysteine–histidine fingers where each finger forms a separate structural unit. Moreover, the multi-cysteine finger cannot be converted to a cysteine–histidine finger by substituting two of its cysteines with histidines (Green and Chambon, 1987). Thus whilst the multi-cysteine domain is clearly similar to the cysteine–histidine domain in its co-ordination of zinc, it is distinct in its lack of histidines and of conserved phenylalanine and leucine residues as well as in its structure, and the two elements are unlikely to be evolutionarily related (for review, see Schwabe and Rhodes, 1991; Rhodes and Klug, 1993).

Whatever its precise relationship to the cysteine–histidine finger, it is clear that, like this type of finger, the multi-cysteine domain in the hormone receptors is involved in mediating DNA binding. Similar single domains containing multiple cysteines separated by non-conserved residues have also been identified in other DNA-binding proteins, such as the yeast transcription factors GAL4, PPRI, LAC9, etc., which all contain a cluster of six invariant cysteines, and in the adenovirus transactivator, E1A, which has a cluster of four cysteines within the region that mediates *trans*-activation (Table 7.2; for review, see Evans and Hollenberg, 1988).

The existence of a short DNA-binding region in a number of different steroid-receptor proteins which bind distinct but related sequences (Section 6.2.3) has allowed a dissection of the elements in this structure that are important in sequence-specific DNA binding (reviewed by Berg,

Table 7.2 Transcriptional regulatory proteins with multiple cysteine fingers

Finger type	Factor	Species
Cys_4Cys_5	Steroid, thyroid receptors	Mammals
Cys_4	E1A	Adenovirus
Cys_6	GAL4, PPRI, LAC9	Yeast

1989). Thus, as illustrated in Table 6.4, the sequences that confer responsiveness to glucocorticoid or estrogen treatment are distinct but related to one another. If the cysteine-rich region of the estrogen receptor is replaced by that of the glucocorticoid receptor, a chimeric receptor is obtained which has the DNA-binding specificity of the glucocorticoid receptor but, because all the other regions of the protein are derived from the estrogen receptor, it continues to bind estrogen. Hence this hybrid receptor induces the expression of glucocorticoid-responsive genes (which carry its DNA-binding site) in response to treatment with estrogen (to which it binds) (Green and Chambon, 1987; Figure 7.16). Further so-called fingerswop experiments using smaller parts of this region have shown that this change in specificity can also be achieved by the exchange of the N-terminal, four-cysteine finger, together with the region immediately following it, which are therefore critical for determining the sequence-specific binding of the DNA.

These findings have been further refined by exchanging individual amino acids in this region of the glucocorticoid receptor for their equivalents in the estrogen receptor. As shown in Figure 7.17, the alteration of the two amino acids between the third and fourth cysteines of the N-terminal finger to their estrogen receptor equivalents results in a glucocorticoid receptor which switches on estrogen-responsive genes. Hence the change of only two critical amino acids within a protein of 777 amino acids can completely change the DNA-binding specificity of the receptor.

Figure 7.16 Effect of exchanging the DNA-binding domain (shaded) of the estrogen receptor with that of the glucocorticoid receptor on the binding of hormone and gene induction by the hybrid receptor.

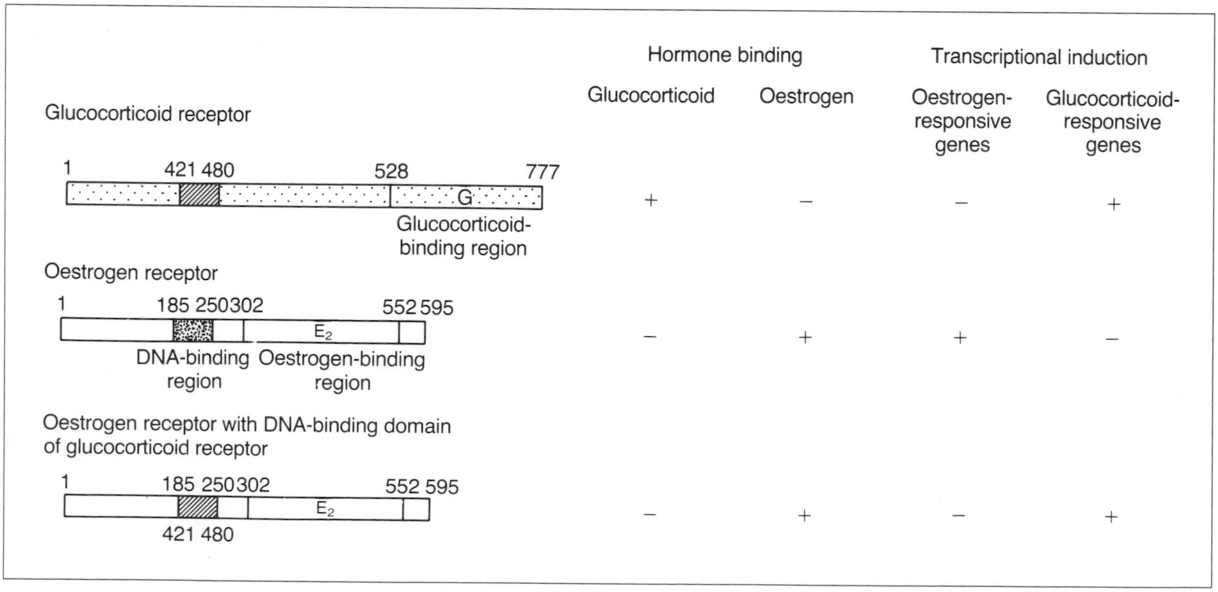

	Hormone binding		Transcriptional induction	
	Glucocorticoid	Oestrogen	Oestrogen-responsive genes	Glucocorticoid-responsive genes
Glucocorticoid receptor	+	–	–	+
Oestrogen receptor	–	+	+	–
Oestrogen receptor with DNA-binding domain of glucocorticoid receptor	–	+	–	+

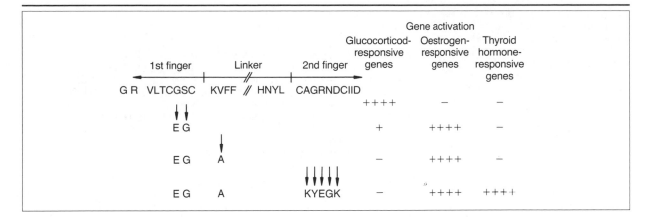

Figure 7.17 Effect of amino acid substitutions in the zinc finger region of the gluco-corticoid receptor on its ability to bind to and activate genes which are normally responsive to different steroid hormones.

The specificity of the hybrid receptor for estrogen-responsive genes can be further enhanced by changing another amino acid, which is located in the linker region between the two fingers (Figure 7.17), indicating that this region also plays a role in controlling the specificity of binding to DNA. Interestingly, this region following the finger can form an α-helical structure similar to the recognition helix seen in the helix-turn-helix motif. Thus, the DNA-binding specificity of the steroid receptors appears to involve the co-operation of a zinc finger motif and an adjacent helical motif.

In contrast to the effect of mutations in the first finger and adjacent region, further alteration of five amino acids in the second finger is sufficient to change the binding specificity of the receptor such that it now recognizes the thyroid hormone receptor binding sites (Umesono and Evans, 1989; Figure 7.17). Since thyroid hormone binding sites do not differ from those of the estrogen receptor in sequence but only in the spacing between the two halves of the palindromic DNA recognition sequence (Table 6.4), this indicates that the second finger is critical for mediating protein–protein interactions between the two copies of the receptor that bind to the two halves of the palindromic sequence (see Section 6.2.3), and thus for controlling the optimal spacing of these halves for binding of the particular receptor.

Hence by studying the multiple related steroid receptors and their relationship with the related DNA sequences to which they bind, it has been possible to determine the critical role of both the first zinc finger and its adjacent helix in controlling the sequence to which these receptors bind, and of the second zinc finger in determining the spacing of adjacent sequences which is optimal for the binding of each receptor. Once again, the steroid-responsive genes are leading the way in the study of gene regulation.

Figure 7.18 Alignment of the leucine-rich region in several cellular transcription factors. Note the conserved leucine residues (L) which occur every seven amino acids.

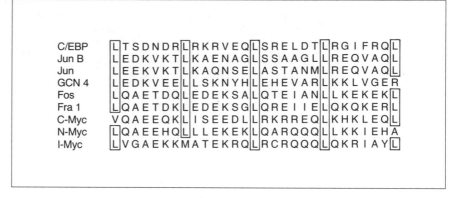

C/EBP	L	T S D N D R	L R K R V E Q	L S R E L D T	L R G I F R Q	L
Jun B	L	E D K V K T	L K A E N A G	L S S A A G L	L R E Q V A Q	L
Jun	L	E E K V K T	L K A Q N S E	L A S T A N M	L R E Q V A Q	L
GCN 4	L	E D K V E E	L L S K N Y H	L E H E V A R	L K K L V G E R
Fos	L	Q A E T D Q	L E D E K S A	L Q T E I A N	L L K E K E K	L
Fra 1	L	Q A E T D K	L E D E K S G	L Q R E I I E	L Q K Q K E R	L
C-Myc	V	Q A E E Q K	L I S E E D L	L R K R R E Q	L K H K L E Q	L
N-Myc	L	Q A E E H Q	L L L L E K E K	L Q A R Q Q Q	L L K K I E H A
I-Myc	L	V G A E K K M A T E K R Q	L R C R Q Q Q	L Q K R I A Y	L

7.2.4 The leucine zipper, the helix-loop-helix motif and the basic DNA-binding domain

In earlier sections of this chapter, we have examined how the presence of unusual structural motifs, such as the zinc finger, in several different regulatory proteins led to the identification of the crucial role of these elements in DNA binding. A similar approach has led to the identification of another such motif, the leucine zipper (for reviews, see Abel and Maniatis, 1989; Lamb and McKnight, 1991). Thus, in studies of the gene encoding the transcription factor C/EBP which is involved in stimulating the expression of several liver-specific genes, it was noted that it contained a region of 35 amino acids, in which every seventh amino acid was a leucine. Similar runs of leucine residues were also noted in the yeast transcriptional regulatory protein GCN4, as well as in the proto-oncogene proteins Myc, Fos and Jun, which were originally identified on the basis of their ability to transform cultured cells to a cancerous phenotype (see Chapter 8) and are believed to act by regulating the transcription of other cellular genes (Figure 7.18).

It was proposed that the leucine-rich region would form an α-helix in which the leucines would occur every two turns on the same side of the helix. These leucine residues would then facilitate the dimerization of two molecules of the transcription factor by promoting the interdigitation of two such helices, one on each of the individual molecules (Figure 7.19). In agreement with this idea, replacement of individual leucine residues in C/EBP with valine or isoleucine residues abolishes the ability of the protein to form a dimer. In turn, these mutations also prevent the binding of the protein to its specific recognition sequence (for review, see Johnson and McKnight, 1989).

Unlike the helix-turn-helix motif, however, the leucine zipper does not bind directly to DNA. Rather, by facilitating the dimerization of the protein it provides the correct protein structure for DNA binding by

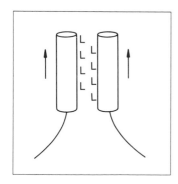

Figure 7.19 Model of the leucine zipper and its role in the dimerization of two molecules of a transcription factor.

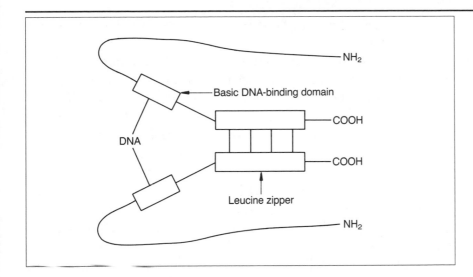

Basic DNA-binding domain

DNA

Leucine zipper

Figure 7.20 Model for the structure of the leucine zipper and the adjacent DNA-binding domain following dimerization of the transcription factor C/EBP.

the adjacent region of the protein, which is rich in basic amino acids that can interact directly with the acidic DNA (Figure 7.20). In agreement with this idea, mutations in the basic DNA-binding domain abolish the ability of the protein to bind to the DNA without abolishing its ability to dimerize. Similar juxtapositions of a basic DNA-binding domain and the leucine zipper are also found in the Fos and Jun oncogene proteins and in the yeast transcription factor GCN4, where a single 60 amino acid region contains all the information needed for both dimerization and sequence-specific DNA binding. Hence the leucine zipper has a role similar to that of the second zinc finger in the steroid receptors (Section 7.2.3) which modulates the activity of the DNA-binding region rather than being involved directly in binding.

Although originally identified in leucine zipper-containing proteins, the basic DNA-binding domain has also been identified by homology comparisons in a number of other transcriptional regulatory proteins which lack the leucine zipper. In this case, however, the basic domain is associated with an adjacent region that can form a helix-loop-helix structure. This motif is distinct from the helix-turn-helix motif described in Section 7.2.2 and consists of two amphipathic helices (containing all the charged amino acids on one side of the helix) separated by an intervening non-helical loop. Although originally thought to be the DNA-binding domain of these proteins, this helix-loop-helix motif is now known to play a similar role to the leucine zipper in mediating protein dimerization and facilitating DNA binding by the adjacent basic DNA-binding motif.

The helix-loop-helix motif with its adjacent basic DNA-binding region is present in a number of different transcription factors expressed in different tissues. It appears, however, to be particularly prevalent in the factors which control muscle development, being found, for example, in the MyoD transcription factor, whose artificial expression in an undifferentiated fibroblast cell line can induce it to differentiate to skeletal muscle cells by activating the expression of muscle-specific genes (for reviews, see Edmondson and Olson, 1993; Buckingham, 1994). Thus, the MyoD gene is likely to be the critical regulatory locus which is activated by treatment of these cells with 5-azacytidine, allowing this agent to induce these cells to differentiate into muscle cells (see Section 5.5.1).

The leucine zipper and helix-loop-helix structures therefore act to mediate the dimerization of the transcription factors which contain them, so forming a dimeric molecule which is able to bind to DNA via the adjacent basic DNA-binding domain. This ability provides an additional aspect to gene regulation by such proteins. Thus, in addition to the formation of a dimer by two identical factors, it is possible to envisage the formation of a heterodimer between two different factors, which might have different properties in terms of sequence-specific binding and gene activation compared to homodimers of one or other of the two factors.

An example of this type is seen in the case of the related oncoproteins Fos and Jun (for review, see Lewin, 1991). Thus Jun can bind as a homodimer to the AP1 recognition sequence, TGAGTCAG, which mediates transcriptional induction by phorbol esters (Table 6.3). In contrast, Fos cannot bind to DNA alone but can form a heterodimer with the Jun protein. This heterodimer binds to the AP1 recognition site with a 30-fold greater affinity than the Jun homodimer and is considerably more effective in enhancing transcription of genes containing the binding site. Both hetero- and homodimer formation and DNA binding are dependent on the leucine zipper motif which is found in both proteins. Hence dimerization by the leucine zipper motif allows two different complexes with different binding affinities and different activity to form on the identical DNA binding site (Figure 7.21).

The failure of the Fos protein to form homodimers and its inability to bind to DNA in the absence of Jun has been shown to be due to differences in its leucine zipper region from that of Jun. Thus if the leucine zipper region of Fos is replaced by that of Jun the resulting protein can dimerize. This dimerization allows the chimeric protein to bind to DNA through the basic region of Fos, which is therefore a fully functional DNA-binding domain.

As well as having a positive role, heterodimerization can also have a negative role. Thus as discussed in Section 7.3.4, the ability of the MyoD factor to stimulate gene expression is inhibited by heterodimerization with

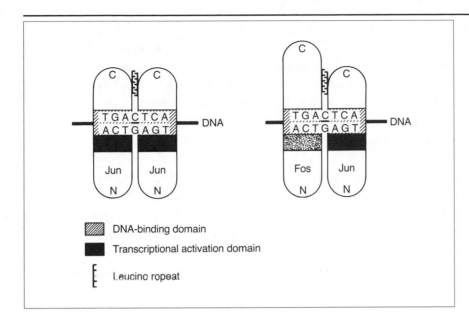

Figure 7.21 Model for DNA binding by the Jun homo-dimer and the Fos–Jun heterodimer.

the Id factor which has a helix-turn-helix motif but no basic domain. Since DNA binding requires the co-operation of two basic domains within the heterodimer, the MyoD–Id heterodimer cannot bind to DNA. Hence MyoD cannot activate the expression of muscle-specific genes and thereby promote the production of skeletal muscle cells in the presence of the Id factor (Figure 7.22). It is clear, therefore, that the ability of the leucine zipper and the helix-loop-helix motif to facilitate the formation of different dimeric complexes between different transcription factors is likely to play a crucial role in the regulation of gene expression, by producing complexes with different binding affinities and different activities (for reviews, see Jones, 1990; Lamb and McKnight, 1991).

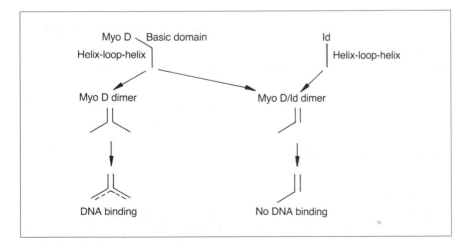

Figure 7.22 DNA binding by the MyoD protein is inhibited by the Id factor which contains the helix-loop-helix dimerization motif but not the basic DNA-binding domain.

7.2.5 Other DNA-binding domains

Although the majority of DNA-binding domains analyzed so far fall into the three classes we have discussed, not all do so. Thus the DNA-binding domains of the transcription factors AP2, CTF/NF1 and SRF are distinct from the known motifs and from each other. As more factors are studied, DNA-binding motifs similar to those of these proteins are likely to be identified and they will become founder members of new families of DNA-binding motifs. Indeed this process is already under way, the similarity between the mammalian factor HNF-3 and the *Drosophila* fork head factor, for example, having led to the identification of a new family of proteins containing the 'so called' fork head DNA binding motif whilst another domain known as the Ets domain has been identified in the *ets*-1 proto-oncogene and several other factors.

It is clear therefore that a number of different structures exist which can mediate sequence-specific DNA binding. Several of these are common to a number of different transcription factors, with differences in the precise amino acid sequence of the motif in each factor controlling the precise DNA sequence which it binds and hence the target genes for the factor.

7.3 Regulation of transcription

7.3.1 Introduction

Although binding to DNA is a necessary prerequisite for the activation of transcription, it is clearly not in itself sufficient for this to occur. Following binding, the bound transcription factor must somehow regulate transcription, e.g. by directly activating the RNA polymerase itself or by facilitating the binding of other transcription factors and the assembly of a stable transcriptional complex. Although some transcription factors can inhibit transcription, the majority of factors act to activate transcription. We will therefore discuss in turn the features of activating transcription factors that produce this activation (Section 7.3.2) and the manner in which they do so (Section 7.3.3), and then discuss the manner in which specific factors inhibit transcription (Section 7.3.4).

7.3.2 Activation domains

Identification of activation domains

It is clear from the preceding sections of this chapter that transcription factors have a modular structure in which a particular region of the protein mediates DNA binding while another may mediate binding of a co-factor, such as a hormone, and so on. It seems likely, therefore, that a specific region of each individual transcription factor will be involved in its ability to up-regulate transcription following DNA binding.

In the majority of cases, it is clear that such activation regions are distinct from those which produce DNA binding. This domain-type structure is seen clearly in the yeast transcription factor GCN4, which mediates the induction of the genes encoding the enzymes of amino acid biosynthesis in response to amino acid starvation. Thus, if a 60 amino acid region of this protein, containing the DNA-binding region, is introduced into cells, it can bind to the DNA of GCN4-responsive genes but fails to activate transcription (Hope and Struhl, 1986). Hence, although DNA binding is necessary for transcriptional activation to occur, it is not sufficient and gene activation must be dependent upon a region of the protein that is distinct from that mediating DNA binding.

Unlike the DNA-binding regions, the region of a transcription factor that mediates gene activation cannot, therefore, be identified on the basis of a simple assay of, for example, the ability to bind to DNA or another protein. Rather, a functional assay of gene activation following binding to DNA is required. Activation regions have therefore been identified on the basis of so-called 'domain-swap' experiments, in which the DNA-binding region of one transcription factor is combined with various regions of another factor and the ability to activate transcription of a gene containing the binding site of the first factor is assessed (Figure 7.23). Following binding of the hybrid factor to the target gene binding site, gene activation will occur only if the hybrid factor also contains an activation domain provided by the second factor, and hence the activation domain can be identified.

Thus, in the case of the yeast transcription factor GCN4 discussed above, if a 60 amino acid region, outside the DNA-binding domain, is linked to the DNA-binding region of the bacterial regulatory protein, Lex A, the hybrid factor will activate transcription in yeast from a gene containing the binding site for Lex A, whereas neither the Lex A DNA-binding domain nor the GCN4 region will do so alone. Hence this region of GCN4 contains an activation domain, which can increase transcription following DNA binding and is separate from the region of the protein that normally mediates DNA binding (Figure 7.24a; Hope and Struhl, 1986).

Following their initial use in yeast, similar domain-swopping experiments have also been used to identify the activation domains of mammalian transcription factors. In the glucocorticoid receptor, for example, a 200 amino acid region at the N-terminus (amino acids 200–400, see

Figure 7.23 Domain-swapping experiment, in which the activation domain of factor 2 is mapped by combining different regions of factor 2 with the DNA-binding domain of factor 1 and assaying the hybrid proteins for the ability to activate transcription of a gene containing the DNA-binding site of factor 1.

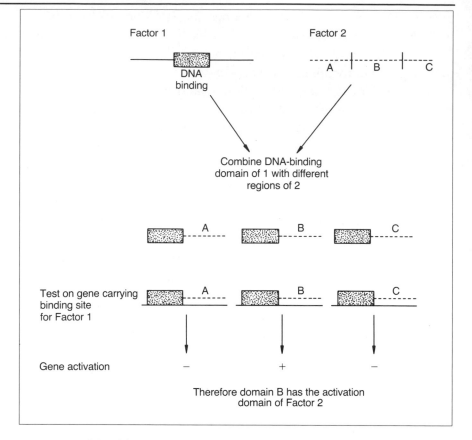

Figure 7.24b), able to produce gene activation, was identified by fusing different regions of the protein to the Lex A DNA-binding domain (Godowski *et al.*, 1988). Similarly, using a slightly different approach, Hollenberg and Evans (1988) found that both this region and a 30 amino acid peptide near the C-terminus of the molecule (amino acids 526–556, see Figure 7.24) produce gene activation independently when fused to the GAL4 DNA-binding domain.

Although other members of the steroid–thyroid hormone receptor family also possess two distinct activation domains, their relative importance in producing gene activation has been shown to vary in different receptors and is also dependent on the target gene that is being activated (reviewed by Green and Chambon, 1988).

The success of domain-swop experiments is further proof of the modular nature of transcription factors, allowing the DNA-binding domain of one factor and the activation domain of another to co-operate together to produce gene activation. This is analogous to the exchange of the DNA-binding domain of the glucocorticoid and estrogen receptors, which allows the creation of a hybrid receptor that binds to estrogen-responsive

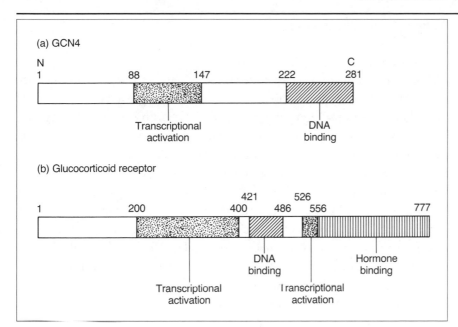

Figure 7.24 Structure of the yeast GCN4 factor (a) and the mammalian gluco-corticoid receptor (b), indicating the distinct regions that mediate DNA binding or transcriptional activation.

genes through the DNA-binding domain but is responsive to the presence of glucocorticoid through the steroid-binding domain of the protein.

An extreme example of this modularity is provided by the herpes simplex virus *trans*-activating protein VP16, which activates transcriptionally the viral immediate-early genes during lytic infection of mammalian cells. Thus, although this protein contains a very potent activation region which can strongly induce gene transcription when fused to the DNA-binding domain of GAL4, it contains no DNA-binding domain and cannot bind to DNA itself. Rather, following infection, it forms a complex with the cellular octamer-binding protein, Oct-1 (see Section 7.2.2). Oct-1 provides the DNA-binding domain which allows binding to the sequence TAATGARAT (R = purine) in the viral promoters, and activation is achieved by the activation domain of VP16 (for review, see Goding and O'Hare, 1989). Hence, in this instance, DNA binding and activation motifs actually reside on separate molecules (Figure 7.25).

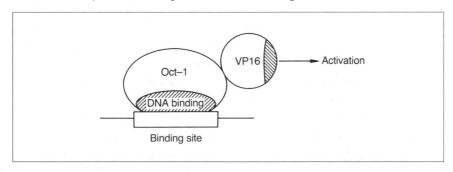

Figure 7.25 Activation of gene transcription by interaction of the cellular factor Oct-l, which contains a DNA-binding domain, and the herpes simplex virus VP16 protein, which contains an activation domain but cannot bind to DNA.

Nature of activating regions

ACIDIC DOMAINS

Although the activating regions identified in various transcription factors do not show strong amino acid sequence homology to each other, in many cases they have a very high proportion of acidic amino acids, resulting in a strong net negative change. Thus, in the N-terminal activating region of the glucocorticoid receptor, 17 acidic amino acids are contained in an 82 amino acid region. Similarly, the activating region of the yeast factor GCN4 contains 17 negative charges in an activating region of only 60 amino acids. This has led to the idea that activation regions consist of so-called 'acid blobs' or 'negative noodles' which are necessary for the activation of transcription (for a review, see Sigler, 1988). Although the net negative change in acidic activation regions is likely to be of importance in their ability to stimulate transcription, different structures such as an α-helix or an antiparallel β-sheet have been suggested for these regions and their precise conformation is at present unclear (for review, see Hahn, 1993b).

OTHER ACTIVATING DOMAINS

Although acidic domains of the type discussed above have been identified in a wide range of transcriptional activators from yeast to man, it is unlikely that this type of structure is the only one which can mediate transcriptional activation. Thus of the two regions of the human Spl transcription factor that could mediate activation of transcription, neither was particularly acidic. Instead, each of these two domains was particularly rich in glutamine residues and the intactness of the glutamine-rich region was essential for transcriptional activation. Similar sequences have also been identified in the homeotic proteins Antennapedia and Cut, in Zeste, another *Drosophila* transcriptional regulator, and in the POU proteins Oct-1 and Oct-2, suggesting that this type of activating region may not be confined to a single protein.

A further type of activation domain has been identified in the transcription factor CTF/NF1 which binds to the CCAAT box present in many eukaryotic promoters (Table 6.2). The activation domain of this protein is not rich in acidic or glutamine residues but instead contains numerous proline residues, forming approximately one-quarter of the amino acids in this region. Similar proline-rich regions are found in other transcription factors, such as AP2 and Jun, suggesting that, as with glutamine-rich domains, this element is not confined to a single protein.

Hence, as with DNA-binding domains, it is clear that several distinct protein motifs are involved in the activation of transcription (for review, see Mitchell and Tjian, 1989).

7.3.3 How is transcription activated?

The widespread interchangeability of activation domains from yeast, *Drosophila* and mammalian transcription factors, discussed in the previous section, suggests that one common mechanism may mediate transcriptional activation in a wide range of organisms. This is supported by the observations that mammalian transcription factors, such as the glucocorticoid receptor, can activate a gene carrying their appropriate DNA-binding site when introduced into yeast cells, while the yeast factor GAL4, which normally activates the transcription of yeast genes in the presence of galactose, can activate a gene carrying its binding site in cells of *Drosophila*, tobacco plants and mammals (reviewed by Guarente, 1988; Ptashne, 1988). Indeed, yeast and mammalian factors can co-operate together in gene activation, a gene bearing binding sites for GAL4 and the glucocorticoid receptor being synergistically activated by the two factors in mammalian cells so that the activation observed with the two factors together is greater than the sum of that observed with each factor independently. This co-operation between two factors which come from widely different species and would therefore never normally interact suggests that they both function by interacting with some highly conserved component of the basal transcriptional complex (see Section 6.2.2) which is involved in the basic process of transcription in all different species.

Clearly, activating factors could stimulate transcription either by increasing the binding of a particular component of the basal transcriptional complex, so enhancing its assembly, or by altering the conformation of an already bound factor, so stimulating the activity and/or stability of the complex. There is evidence that both of these mechanisms are actually used and they will be discussed in turn.

Factor binding

As described in Section 6.2.2, the binding of the TFIID complex (containing TBP and associated proteins) to the TATA box is followed by the binding of the TFIIB factor which in turn promotes binding of RNA polymerase and associated factors. It has been shown (Choy and Green, 1993) that the binding of TFIIB is greatly stimulated in the presence of either acidic or non-acidic activators compared with its low rate of binding in the absence of such an activator. In contrast the rate of binding of TFIID is unchanged whether or not activators are present. Hence

Figure 7.26 The binding of an activator (A) to its appropriate binding site (ABS) can enhance the rate of transcription by enhancing the binding of the TFIIB factor which in turn promotes assembly of the basal transcriptional complex.

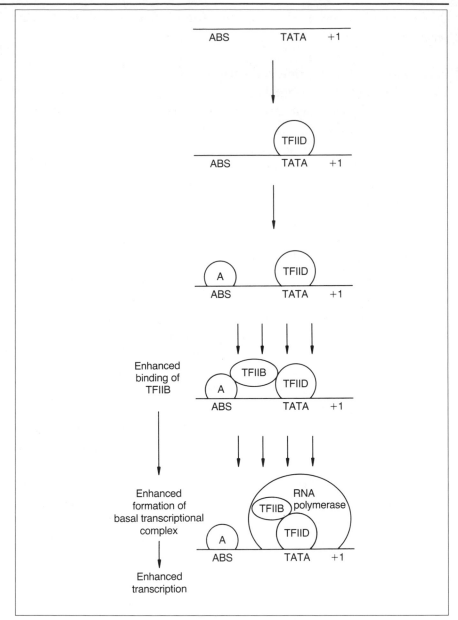

activators can stimulate the assembly of the basal transcriptional complex by enhancing the recruitment of TFIIB (Figure 7.26).

As envisaged in the model shown in Figure 7.26, there is evidence that acidic activators interact directly with TFIIB. Thus TFIIB can be purified from a mixture of proteins on a column containing a bound acidic activator. Moreover, this interaction is of importance for activation of transcription since mutations in acidic activators which abolish the interaction with

TFIIB prevent the activator from stimulating transcription (for reviews, see Sharp, 1991; Hahn, 1993a). Hence the ability of activators to enhance the recruitment of TFIIB and thereby promote assembly of the basal transcriptional complex is an essential component of their ability to stimulate transcription.

Factor activity

The stimulation of factor binding and complex assembly is not the only mechanism by which activators can stimulate transcription. Thus activators also appear to act at a second point following the initial assembly of the complex (Choy and Green, 1993). This could involve enhancing the stability of the complex or stimulating its activity, so increasing the likelihood that it will initiate transcription (Hahn, 1993a; Figure 7.27).

Although it is generally accepted that activators act at this second step, the target molecule in the basal transcriptional complex with which they interact in order to do this remains unclear. One obvious possibility in this regard is TFIIB itself since, as described above, it clearly does interact with

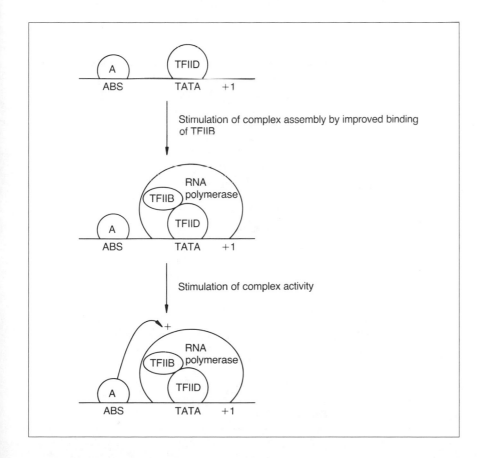

Figure 7.27 An activator (A) bound to its binding site (ABS) can stimulate complex assembly at a stage after improved binding of TFIIB. This appears to occur via the interaction of the factor with the basal transcriptional complex containing RNA polymerase, TFIIB, TFIID and other factors.

activating molecules and could therefore also be involved at this second stage. However, it is possible that the interaction with TFIIB is confined to the initial stage of complex assembly and that other factors are the targets at the second stage.

One obvious target would be the RNA polymerase enzyme itself, so allowing an activator to directly stimulate enzyme activity (Figure 7.28a). Thus the largest subunit of RNA polymerase II is known to contain a repeating motif of a sequence whose consensus is Tyr-Ser-Pro-Thr-Ser-Pro-Ser, which is conserved from yeast to mammals. This element is very rich in hydroxyl groups, which could potentially interact with the negative domain of the activation region (for discussion, see Sigler, 1988), while the conservation of this region would explain why yeast activation regions work in mammalian cells and vice versa. Hydrogen bonding to this region of the RNA polymerase could also occur via the amide groups in the glutamine-rich region of the Spl activation domain.

Although this direct mechanism of activation is certainly attractive, no direct evidence in its favor has thus far been obtained and it is more probable that activators interact indirectly with the polymerase via another transcription factor (Figure 7.28b). One obvious candidate in this regard is TFIID.

Direct evidence for an effect of activator molecules on TFIID has been obtained by studying the interaction of TFIID and the yeast transcription factor GAL4 on a promoter containing both a TATA box and binding sites

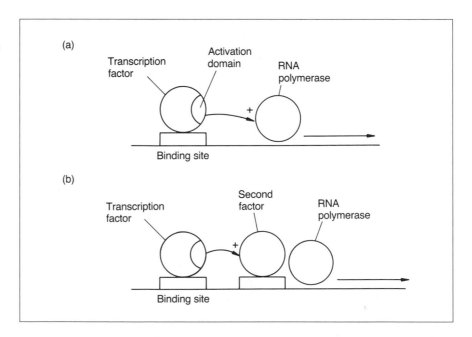

Figure 7.28 Models for the action of an activation domain either directly on the RNA polymerase (a) or via a second factor (b).

for GAL4 (Horikoshi *et al.*, 1988). Although the level of binding of TFIID to the TATA box is not altered in the presence of an acidic activator, its conformation is altered by addition of such an activator. Thus in the presence of the activating form of the yeast transcription factor GAL4, TFIID binding was detected over both the TATA box and the start site for transcription, whereas in its absence TFIID was bound only at the TATA box site (Figure 7.29). In contrast, no change in TFIID binding was observed in the presence of a form of GAL4 containing the DNA-binding domain but lacking the acidic region necessary for activation. Hence the change in TFIID binding is directly correlated with the ability of GAL4 to activate transcription and is not simply a consequence of DNA binding.

Similar direct interaction with TFIID has also been observed for the mammalian transcription factor ATF, which mediates the increased transcription of several cellular genes following cyclic AMP treatment, suggesting that this is a general mechanism by which activation domains function (Figure 7.30).

Thus activators are capable of interacting with TFIID to alter its configuration and such a change presumably enhances either the stability or the activity of the basal transcriptional complex. The idea that activators interact directly with TFIID is supported by studies which show that like TFIIB, TBP (the DNA-binding component of TFIID) can be purified on a column containing a bound activator whilst mutations which prevent the activator binding to TBP abolish its ability to stimulate transcription.

Although there is therefore evidence that activators can interact directly with the TBP component of TFIID (Figure 7.31a), other evidence has

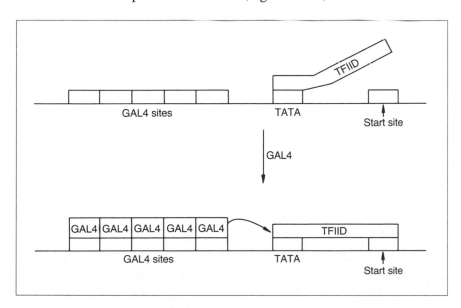

Figure 7.29 Effect of GAL4 binding on the binding of TFIID.

Figure 7.30 Binding of ATF to its binding site alters the binding of TFIID and facilitates the assembly of a stable transcription complex.

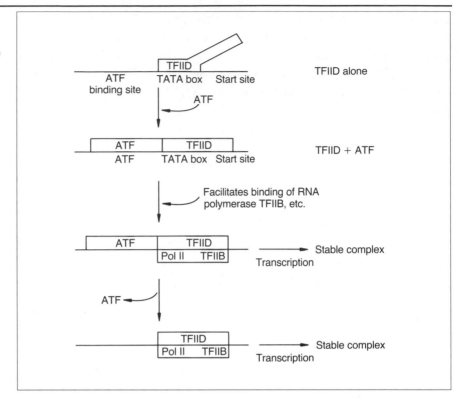

suggested that some of the effects of activating molecules can only be observed in the presence of the entire TFIID complex and not of purified TBP. This has led to the idea that activators can also act on TBP indirectly via other components of the TFIID complex. These so called 'adaptor molecules' would not stimulate transcription themselves but would link activators with TBP which in turn would interact with RNA polymerase (Figure 7.31b) (for reviews, see Lewin, 1990; Gill and Tjian, 1992).

There exists therefore a bewildering array of potential targets for activator molecules, including the RNA polymerase itself, TFIIB and different components of TFIID. These possibilities are not mutually exclusive however. Thus it is likely that all these components can act as a target either for the same activating factor or different activating factors. This multiplicity of targets would account for the strong synergistic activation of transcription which is observed when different activating factors are added together compared to the level observed when each factor is added separately (see above) as well as the strong enhancement of transcription observed when multiple copies of a single factor can bind the target DNA. Thus different activating factors or different molecules of the same activating factor could contact different targets within the basal transcriptional complex, so ensuring the great enhancement of transcriptional activity

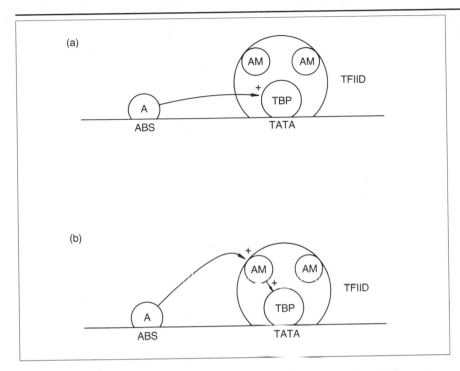

Figure 7.31 An activating molecule (A) can interact with TFIID either by interacting with the TATA box binding protein (TBP) directly (a) or by interacting indirectly with TBP via accessory molecules (AM) (b).

which is the ultimate aim of activating molecules (Figure 7.32) (for review, see Tjian and Maniatis, 1994).

7.3.4 Repression of transcription

Although the majority of transcription factors described so far act in a positive manner, a number of cases have now been reported in which a transcription factor exerts an inhibiting effect on transcription and several possible mechanisms by which this can be achieved have been described (for reviews, see Jackson, 1991; Clark and Docherty, 1993; Figure 7.33).

The simplest means of achieving repression is seen in the β-interferon promoter where the binding of two positively acting factors is necessary for gene activation. Another factor acts negatively by binding to this

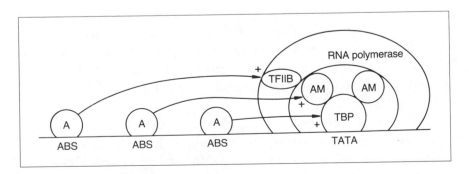

Figure 7.32 Synergistic activation of transcription by activator molecules (A) interacting with different components of the basal transcriptional complex such as TFIIB or the different components of TFIID (TBP and non-DNA-binding accessory molecules).

Figure 7.33 Mechanisms by which an inhibitory factor (R) can repress transcription. These involve inhibiting an activator binding to DNA either by competing for the DNA binding site (a) or by sequestering the activator in solution (b), inhibiting the abllity of bound activator to stimulate transcription (c) or direct repression (d).

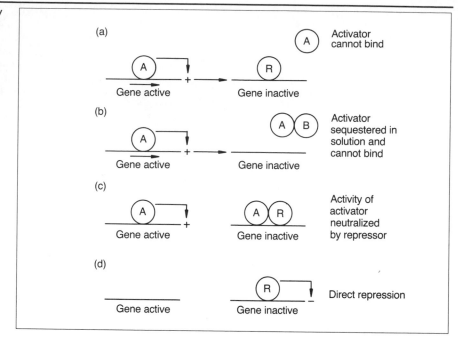

region of DNA and simply preventing the positively acting factors from binding (Figure 7.33a). In response to viral infection, the negative factor is inactivated, allowing the positively acting factors to bind and transcription occurs. A similar example was described in Section 7.2.2, whereby the DNA binding of the engrailed gene product inhibits gene activation by preventing the binding of the Ftz activator protein. In a related phenomenon (Figure 7.33b) repression is achieved by formation of a complex between the activator and the repressor in solution, preventing the activator binding to the DNA. This is seen in the case of the inhibitory factor Id which dimerizes with the muscle-determining factor MyoD via its helix-loop-helix motif and prevents it binding to DNA and activating transcription (see Section 7.2.4).

In addition to inhibiting DNA binding, a negative factor can also act by interfering with the activation of transcription mediated by a bound factor in a phenomenon known as quenching (Figure 7.33c). Thus the negatively acting yeast protein GAL80 prevents gene activation by the GAL4 protein by binding to it and masking the activation domain of GAL4. In response to the presence of galactose, however, GAL80 dissociates allowing GAL4 to activate the genes required for the metabolism of galactose (Figure 7.34).

In all cases discussed so far the negative factor exerts its inhibiting effect in an essentially passive manner by neutralizing the action of a positively acting factor by preventing either its DNA binding or its activation of

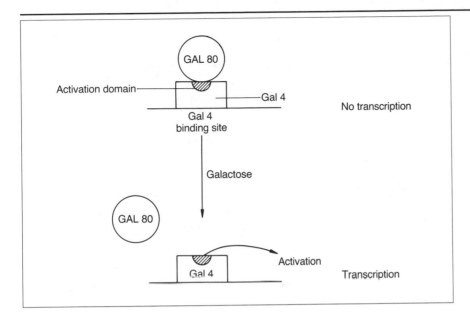

Figure 7.34 Galactose activates gene expression by removing the GAL80 protein from the DNA-bound GAL4 protein, unmasking the activation domain on GAL4.

transcription. It has recently become clear, however, that some factors can have an inherently negative action on transcription which does not depend upon the neutralization of a positively acting factor (Figure 7.33d).

Thus the *Drosophila* even-skipped protein is able to repress transcription from a promoter lacking DNA binding sites for any activating proteins, indicating that it has a directly negative effect on transcription (Han and Manley, 1993). A similar direct inhibitory effect has been defined in the case of the mammalian c-*erbA* gene which encodes the thyroid hormone receptor, a member of the steroid–thyroid hormone receptor family (Sections 6.2.3 and 7.2.2). Thus the binding of this receptor to its specific DNA binding site in the absence of thyroid hormone results in the direct inhibition of transcription (Baniahmad *et al.*, 1992). In the presence of thyroid hormone, however, the receptor undergoes a conformational change which presumably unmasks its activation domain and now allows it to activate rather than repress transcription (Figure 7.35). This phenomenon indicates that the distinction between activators and repressors is not a precise one. Thus in some cases the same factor can directly stimulate or directly inhibit transcription depending on the circumstances.

In both the even-skipped and thyroid hormone receptor cases, a specific region of the protein has been shown to be able to confer the ability to inhibit gene expression upon the DNA-binding domain of another protein when the two are artificially linked (Baniahmad *et al.*, 1992; Han

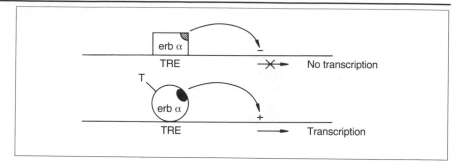

Figure 7.35 The thyroid hormone receptor encoded by the c-*erbA* α gene represses transcription in the absence of thyroid hormone (T) via a specific inhibitory domain (hatched). Following binding of hormone (T), the protein undergoes a confor-mational change which exposes its activation domain (solid) and allows it to activate transcription.

and Manley, 1993), paralleling the similar behavior of activation domains in such domain-swop experiments (Section 7.3.2). By analogy with acti-vation domains such inhibitory domains are likely to inhibit transcription by reducing the formation and/or stability of the basal transcriptional complex.

Hence inhibitory factors are likely to play a critical role in the regulation of transcription both by inhibiting the activity of positively acting factors and in some cases by having a direct inhibitory effect on transcription.

7.4 What regulates the regulators?

7.4.1 Introduction

The crucial role of transcription factors in tissue-specific gene regulation and in the regulation of transcription in response to specific stimuli clearly leads to the question of how such factors are themselves regulated so that they produce gene activation or repression only in a specific tissue or in response to a particular signal.

As discussed in Chapter 6 (Section 6.1.3), the Britten and Davidson model envisaged that the products of regulatory or integrator genes which controlled the activity of other genes would be synthesized in response to a particular signal, or in a particular tissue, and would be absent in other tissues. Hence the activity of the regulated genes would be directly corre-lated with the presence of the regulatory protein (Figure 7.36a).

An example of this type is provided by the octamer-binding protein Oct-2 (see Section 7.2.2), which is involved in the stimulation of immunoglobulin gene expression in B cells. In agreement with this, the Oct-2 protein and the RNA encoding it are present in B cells but are absent in other cell types, such as HeLa cells, which do not express the immunoglobulin genes. Moreover, artificial expression of the gene encod-ing Oct-2 in HeLa cells results in the transcription of immunoglobulin

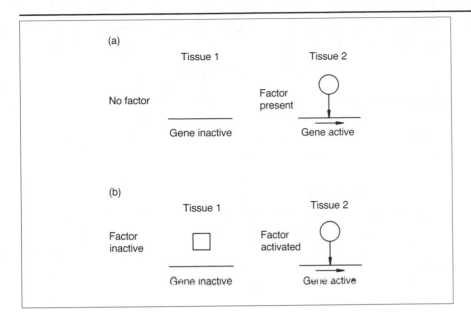

Figure 7.36 Gene activation mediated by the synthesis of a transcription factor only in a specific tissue (a) or its activation in a specific tissue (b).

genes introduced into these cells, confirming that Oct-2 induces transcription of immunoglobulin genes in tissues that contain it (Muller *et al.*, 1988).

A contrasting example is provided, however, by another factor involved in immunoglobulin gene transcription, i.e. NF κB. This factor is detected in a form capable of stimulating immunoglobulin gene transcription only in extracts of mature immunoglobulin-producing B cells, no activity being detectable in immature B cells or in non-B cells, such as HeLa cells. The NF κB protein and its corresponding RNA are detectable in all cell types, however, suggesting that this protein exists in an inactive form in most cell types and is activated in mature B cells. In agreement with this, active NF κB capable of stimulating immunoglobulin gene transcription can be induced in other cell types, such as T cells or HeLa cells, by treatment with phorbol esters. This effect occurs even under conditions where new protein synthesis cannot occur, indicating that it takes place via the activation of pre-existing NF κB protein (reviewed by Grimm and Baeurele, 1993).

These two examples illustrate, therefore, how the ability of transcription factors to act only in a particular tissue can be controlled either by tissue-specific synthesis (Figure 7.36a) or by tissue-specific activation of pre-existing protein (Figure 7.36b). These two mechanisms will now be considered.

7.4.2 Regulated synthesis of transcription factors
Clearly, all the various levels of gene regulation such as transcription, splicing, translation, etc., which were discussed in Chapter 3 (Section 3.1),

could be used to regulate the expression of the genes encoding transcription factors and there is evidence that several of these are used. These will be discussed in turn with illustrative examples.

Regulation of transcription

Although the low abundance of many transcription factors makes the demonstration of transcriptional control difficult, it has been demonstrated in the case of the C/EBP protein which, as previously described (Section 7.2.4), regulates the transcription of several different liver-specific genes, such as transthyretin and α-l-antitrypsin. This transcription factor is made at high level only in the liver and, in agreement with this, significant transcription of its gene is detectable only in this tissue (Xanthopoulos *et al.*, 1989). Hence the regulated transcription of the C/EBP gene controls the production of the corresponding protein which, in turn, directly controls the liver-specific transcription of other genes.

Regulation of splicing

As discussed in Chapter 4 (Section 4.2.3), in mammals alternative splicing is used widely to generate two different forms of a protein, with different functions, from a single gene. A similar theme is seen in the transcription factors, and has been well characterized in the c-*erbA* α gene, which encodes a receptor mediating gene expression in response to thyroid hormone. As discussed in Section 7.3.4, gene activation by this protein is dependent on the presence of thyroid hormone, and in the absence of hormone the receptor represses gene expression. Activation by the receptor is dependent upon the presence of a region in the receptor capable of binding thyroid hormone. Interestingly, however, two alternative forms of the protein exist which are encoded by alternatively spliced mRNAs. One of these (α-1) contains the hormone-binding domain and can mediate gene activation in response to thyroid hormone, while the other form (α-2) contains another protein sequence instead of part of this domain (Figure 7.37a). Therefore the α-2 form cannot respond to the hormone, although since it contains the DNA-binding domain, it can bind to the binding site for the receptor in hormone-responsive genes. By doing so, it prevents binding of the α-1 form and hence the induction of the gene in response to thyroid hormone (Figure 7.37b; for review, see Foulkes and Sassone-Corsi, 1992). These two alternatively spliced forms of the transcription factor, which are made in different amounts in different tissues, therefore mediate opposing effects on thyroid-hormone-dependent gene expression.

Similar alternative splicing events which produce activators or repres-

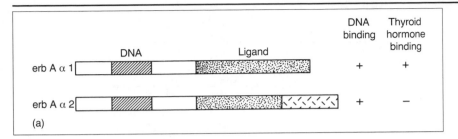

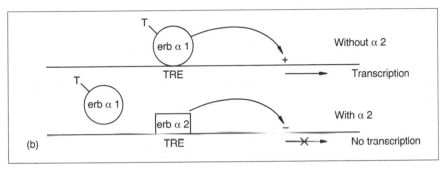

Figure 7.37 (a) Relationship of the ErbA α-l and α-2 proteins. Note that only the α-l protein has a functional thyroid-hormone-binding domain. (b) Inhibition of ErbA α-1 binding and of gene activation in the presence of the α-2 protein.

sors from the same gene have been described in a number of other cases (Foulkes and Sassone-Corsi, 1992). Thus, for example, the Oct-2 transcription factor, which is a member of the POU family (Section 7.2.2), undergoes cell-type-specific alternative splicing in B lymphocytes and neuronal cells. The predominant form produced in B cells has a C-terminal activation domain and can therefore stimulate gene expression following DNA binding. In contrast, the predominant neuronal form lacks this region, and can therefore bind to DNA and inhibit gene activation by other octamer-binding proteins (Lillycrop and Latchman, 1992).

Regulation of translation

As discussed in Chapter 4 (Section 4.5.2), the yeast transcription factor GCN4, which induces the expression of genes involved in amino acid biosynthesis in response to amino acid starvation, is itself regulated at the level of translation (reviewed by Fink, 1986). Thus, when amino acids are lacking, translation is initiated preferentially at the start point for GCN4 production, whereas when amino acids are abundant, translation initiates at short open reading frames upstream of the GCN4 start site and does not reinitiate at the GCN4 site (Figure 4.24). Hence GCN4 is synthesized in response to amino acid starvation and activates the genes encoding the enzymes required for the biosynthetic pathways necessary to make good this deficiency.

Such regulation of translation by alternative initiation is also observed in mammalian cells in the case of a transcription factor expressed in the

Figure 7.38 Alternative use of different initiation codons results in the generation of the LAP protein which can activate transcription due to its possession of a functional activation domain and the LIP protein which lacks this domain and therefore acts as a repressor.

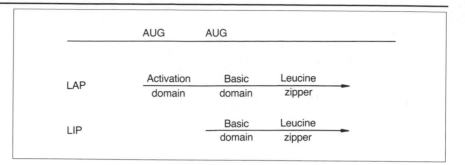

liver (see Section 4.5.2). In this case, however, alternative initiation produces two different proteins, liver activator protein (LAP) and liver inhibitor protein (LIP). Only LAP contains the N-terminal activation domain. In contrast LIP production involves translational initiation at a downstream AUG residue so that this protein lacks the activation domain and contains only the DNA-binding domain and the leucine zipper regions which are also found in LAP (Figure 7.38). LIP is therefore capable of dimerizing and binding to DNA, but obviously cannot activate transcription (Figure 7.38). It therefore acts as an inhibitory protein which interferes with the activation of transcription by LAP by competing with it for binding to the DNA (Descombes and Schibler, 1991). This use of alternative initiation codons to produce different forms of a transcription factor with opposite effects on transcription thus parallels the similar use of alternative splicing to do this in factors such as the thyroid hormone receptor (see above).

7.4.3 Regulated activity of transcription factors

Although, as discussed above, many transcription factors are regulated by controlling their synthesis, a number are controlled by regulating the activity of the protein, which is present in many different cell types. Such a system has obvious advantages in that it allows a direct effect of the agent inducing gene expression on the activity of the factor, either by binding to it or by modifying the protein, for example by phosphorylation, and hence results in a rapid response. Such post-translational modifications therefore allow specific signal transduction pathways to alter the activity of cellular transcription factors and so produce alterations in gene expression (for reviews, see Karin, 1992; Karin and Smeal, 1992). Examples of these kinds of modifications will now be discussed (Figure 7.39)

Activation by protein–ligand or protein–protein interaction
A simple example of a protein–ligand interaction activating a transcription factor is provided by the ACEl factor in yeast, which mediates the induc-

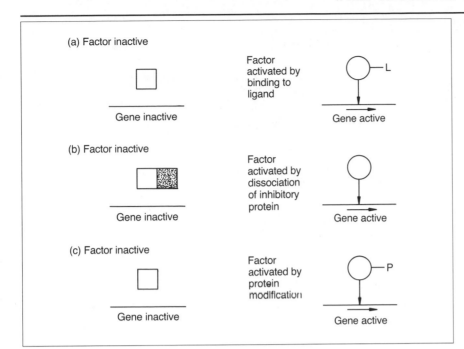

(a) Factor inactive

Gene inactive

Factor activated by binding to ligand

Gene active

(b) Factor inactive

Gene inactive

Factor activated by dissociation of inhibitory protein

Gene active

(c) Factor inactive

Gene inactive

Factor activated by protein modification

Gene active

Figure 7.39 Mechanisms by which transcription factors can be activated by post-translational changes.

tion of the metallothionein gene in response to copper. This protein has been shown to undergo a major conformational change upon binding of copper, which allows it to bind to metallothionein gene regulatory sites and induce transcription (Furst *et al.*, 1988). Hence the activity of this transcription factor is directly modulated by copper, allowing it to mediate gene activation in response to the metal (Figure 7.40). A similar case of a ligand-induced activation is seen with the thyroid hormone receptor (see Section 7.3.4). In this case, however, rather than affecting the ability of the receptor to bind to DNA, the ligand induces a conformational change in the receptor which converts it from a repressor to an activator and hence allows it to activate gene expression in response to thyroid hormone treatment (Figure 7.35).

In contrast, other members of the steroid–thyroid hormone receptor family do not bind to DNA in the absence of the appropriate hormone. This is seen in the case of the glucocorticoid receptor, for example, where the receptor is normally located in the cytoplasm, and only binds to DNA and activates transcription when the hormone is added. Originally it was thought that, as with ACEl, binding of hormone to the receptor activated its ability to bind to DNA and switch on transcription of hormone-responsive genes. However, more recently it has been shown that although the receptor binds to DNA only in the presence of the hormone in the cell, in the test tube it will bind even when no hormone is present. This has led

Figure 7.40 Activation of the ACEI factor in response to copper results in transcription of the metallothionein gene.

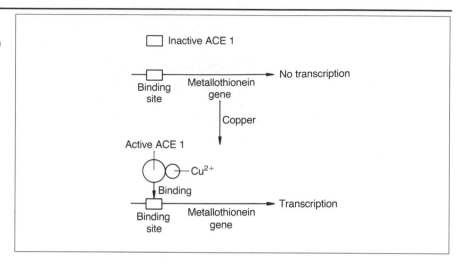

to the idea that in the cell the receptor is prevented from binding to DNA by its association with another protein, and that the hormone acts to release it from this association and allow it to fulfil its inherent ability to bind to DNA.

In agreement with this idea, the glucocorticoid receptor, unlike the thyroid hormone receptor, has been shown to be associated in the cytoplasm with a 90 000 molecular weight heat-inducible protein (hsp90). Upon steroid binding the receptor dissociates from the hsp90 and moves to the nucleus, where it activates gene transcription (Figure 7.41). Hence the transcription factor is activated by hormone not by a protein–ligand interaction but by disruption of a protein–protein interaction which inhibits the inherent DNA-binding ability of the receptor.

Protein–protein interaction also inhibits the activity of the yeast GAL4 transcription factor in the absence of galactose. This interaction, however, still allows the factor to bind to DNA but its acidic activating region is masked by another protein, GAL80 (see Section 7.3.4). However, in the presence of galactose, GAL80 dissociates and activation can occur (Figure 7.34). Interestingly, if an acidic region is linked artificially to GAL80, it will activate transcription through the DNA-bound GAL4 protein in the same way as the herpes simplex virion protein VP16 acts through bound octamer-binding protein (see Section 7.3.2).

Activation by protein modification
Many transcription factors are modified extensively by the addition, for example, of *O*-linked monosaccharide residues or by phosphorylation (for reviews, see Hunter and Karin, 1992; Jackson, 1992). Such modifications represent obvious targets for agents that induce gene activation. Thus,

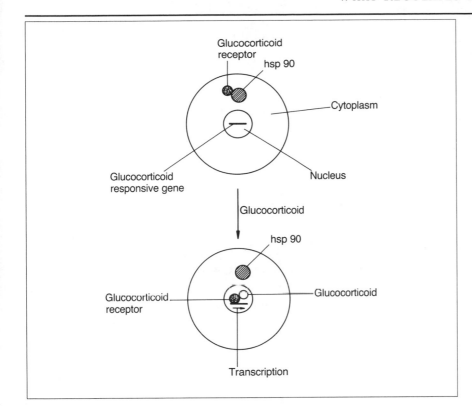

Glucocorticoid receptor

hsp 90

Cytoplasm

Glucocorticoid responsive gene

Nucleus

Glucocorticoid

hsp 90

Glucocorticoid receptor

Glucocorticoid

Transcription

Figure 7.41 Binding of glucocorticoid to the glucocorticoid receptor results in its dissociation from hsp90 and movement of the hormone–receptor complex to the nucleus where it activates the transcription of glucocorticoid-responsive genes.

such agents could act by altering the activity of a modifying enzyme, such as a kinase, which in turn would act on the transcription factor, resulting in its activation and the switching on of gene expression.

The best-characterized example of this type involves the mammalian CREB factor, which mediates the induction of several genes in response to treatment with cyclic AMP and binds to the cyclic AMP response element in these genes (Table 6.3). The ability of CREB to stimulate transcription is enhanced greatly by its phosphorylation by protein kinase A on the serine residue located at amino acid 133 of the protein. Such phosphorylation stimulates the activity of the activation domain located in this region of the protein. Hence the activation of gene expression by cyclic AMP is mediated by its activation of protein kinase A which, in turn, phosphorylates and activates CREB (Figure 7.42) (for review, see Karin and Smeal, 1992).

The regulation of phosphorylation is also likely to be involved in the activation of NF κB in response to phorbol ester treatment, as described above (Section 7.4.1). In this case, however, the target for phosphorylation is not the transcription factor itself but rather an associated protein, I κB, which inhibits NF κB activity. Prior to phorbol ester treatment, NF κB is bound to I κB as an inactive complex which is located in the cytoplasm.

Figure 7.42 Phosphorylation of the CREB transcription factor in response to cyclic AMP stimulates the ability of CREB to activate transcription. It therefore results in the activation of cyclic-AMP-inducible genes containing the cyclic AMP response element to which CREB binds.

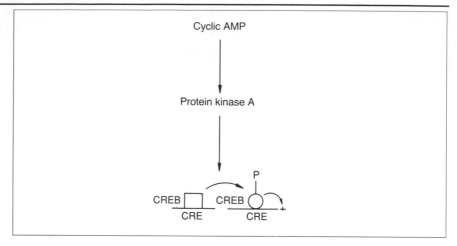

Following treatment, however, the complex dissociates and NF κB moves to the nucleus where it activates gene expression (reviewed by Grimm and Baeurele, 1993). Such activation is clearly similar to the dissociation of the glucocorticoid receptor from hsp90 following steroid treatment, although in this case phosphorylation rather than steroid binding dissociates the protein–protein complex. This example therefore links the two types of activation processes we have discussed, involving both protein–protein interaction and reversible protein modification.

7.5 Conclusions

Although the existence of transcription factors had been known or inferred for some time, it is clear that the cloning of the genes encoding these factors has increased our knowledge of their functional properties greatly. A number of different motifs which mediate DNA binding, protein dimerization and transcriptional activation have been identified, and these are listed in Table 7.3. The list is likely to grow, however, as more factors are cloned and characterized.

Further information is also likely to become available about the manner in which transcription factors become activated in particular tissues or in response to particular stimuli. Already it is clear, however, that unlike most cases of gene regulation, the regulation of transcription factors is often modulated by post-transcriptional processes, including protein modification. Such post-transcriptional regulation is sensible in that transcriptional regulation of the genes encoding these factors would only set the process one step back, leading to the problem of how these genes are themselves

Table 7.3 Transcription factor domains

Domain	Role	Factors containing domain	Comments
Homeobox	DNA binding	Numerous *Drosophila* homeotic genes, related genes in other organisms	DNA binding mediated via helix-turn-helix motif
Cysteine–histidine zinc finger	DNA binding	TFIIIA, Kruppel, SP1, etc.	Multiple copies of finger motif
Cysteine–cysteine zinc finger	DNA binding	Steroid–thyroid hormone receptor family	Single pairs of fingers, related motifs in Adenovirus E1A and yeast GAL4, etc.
Basic element	DNA binding	C/EBP, *c-fos*, *c-jun*, GCN4	Often found in association with leucine zipper
Leucine zipper	Protein dimerization	C/EBP, *c-fos*, *c-jun*, GCN4, *c-myc*	Mediates dimerization which is essential for DNA binding by adjacent domain
Helix-loop-helix	Protein dimerization	*c-myc*, *Drosophila* daughterless, MyoD, E12, E47	Mediates dimerization which is essential for DNA binding by adjacent domain
Amphipathic α-helix	Gene activation	Yeast GCN4, GAL4, steroid–thyroid receptors, etc.	Probably interacts directly with TFIID
Glutamine-rich region	Gene activation	SP1	Related regions in Oct-1, Oct-2, AP2, etc.
Proline-rich region	Gene activation	CTF/NF1	Related regions in AP2, *c-jun*, Oct-2

transcriptionally regulated. In contrast, regulation at the post-transcriptional level allows a direct response to particular stimuli and results, for example, in the rapid increase in the translation of the GCN4 protein in response to amino acid starvation, the rapid activation of the glucocorticoid receptor by dissociation from hsp90 in response to a hormone, and the rapid activation of CREB by phosphorylation in response to cyclic AMP.

Our understanding of the functional domains of these factors and of their activation now opens the way for the detailed molecular modeling of how these factors interact with each other and the RNA polymerase itself, and an elucidation of the way in which they stimulate transcription.

References

Abel, T. and T. Maniatis 1989. Action of leucine zippers. *Nature* **341**, 24–5.

Affolter, M., A. Schier and W. J. Gehring 1990. Homeodomain proteins and the regulation of gene expression. *Current Opinion in Cell Biology* **2**, 485–95.

Akam, M. 1989. Hox and HOM: homologous gene clusters in insects and vertebrates. *Cell* **57**, 347–9.

Baniahmad, A., A. C. Kohne and R. Renkawitz 1992. A transferable silencing domain is present in the thyroid hormone receptor in the v-*erbA* oncogene product and in the retinoic acid receptors. *EMBO Journal* **11**, 1015–23.

Beato, M. 1989. Gene regulation by steroid hormones. *Cell* **56**, 335–44.

Berg, J. M. 1989. DNA binding specificity of steroid receptors. *Cell* **57**, 1065–8.

Buckingham, M. 1994. Which myogenic factors make muscle? *Current Biology* **4**, 61–3.

Choy, B. and M. R. Green 1993. Eukaryotic activators function during multiple steps of preinitiation complex assembly. *Nature* **366**, 531–6.

Clark, A. R. and K. Docherty 1993. Negative regulation of transcription in eukaryotes. *Biochemical Journal* **296**, 521–41.

Descombes, P. and U. Schibler 1991. A liver enriched transcriptional activator protein LAP and a transcriptional inhibitory protein LIP are translated from the same mRNA. *Cell* **67**, 569–80.

Edelman, G. M. and F. S. Jones 1993. Outside and downstream of the homeobox. *Journal of Biological Chemistry* **268**, 20683–6.

Edmondson, D. G. and E. N. Olson 1993. Helix-loop-helix proteins as regulators of muscle-specific transcription. *Journal of Biological Chemistry* **268**, 755–8.

Evans, R. M. 1988. The steroid and thyroid hormone receptor gene superfamily. *Science* **240**, 889–95.

Evans, R. M. and S. M. Hollenberg 1988. Zinc fingers: gilt by association. *Cell* **52**, 1–3.

Fink, G. R. 1986. Translational control of transcription in eukaryotes. *Cell* 45, 155–6.

Foulkes, N. S. and P. Sassone-Corsi 1992. More is better; activator and repressors from the same gene. *Cell* 68, 411–14.

Furst, P., S. Hu, R. Hackett and D. Hamer 1988. Copper activates methallothionein gene transcription by altering the conformation of a specific DNA binding protein. *Cell* 55, 705–17.

Gehring, W. J. 1987. Homeoboxes in the study of development. *Science* 236, 1245–52.

Gill, G. and R. Tjian 1992. Eukaryotic coactivators associated with the TATA box binding protein. *Current Opinion in Genetics and Development* 2, 236–42.

Goding, C. and P. O'Hare 1989. Herpes simplex virus Vmw65-octamer binding protein interaction: a paradigm for combinational control of transcription. *Virology* 173, 363–7.

Godowski, P. J., D. Picard and K. Yamamoto 1988. Signal transduction and transcriptional regulation by glucocorticoid receptor–Lex A fusion proteins. *Science* 241, 812–16.

Green, S. and P. Chambon 1987. Oestradiol induction of a glucocorticoid response gene by a chimaeric receptor. *Nature* 325, 75–8.

Green, S. and P. Chambon 1988. Nuclear receptors enhance our understanding of transcription regulation. *Trends in Genetics* 4, 309–14.

Grimm, S. and P. A. Baeurele 1993. The inducible transcription factor NF-KappaB. *Biochemical Journal* 290, 309–12.

Guarente, L. 1988. UASs and enhancers: common mechanism of transcriptional activation in yeast and mammals. *Cell* 52, 303–5.

Hahn, S. 1993a. Efficiency in activation. *Nature* 363, 672–3.

Hahn, S. 1993b. Structure (?) and function of acidic transcription activators. *Cell* 72, 481–5.

Han, K. and J. L. Manley 1993. Transcriptional repression by the *Drosophila* even-skipped protein: definition of a minimal repression domain. *Genes and Development* 7, 491–503.

Hanes, S. D. and R. Brent 1989. DNA specificity of the bicoid activator protein is determined by homeodomain recognition helix 9. *Cell* 57, 1275–83.

Harrison, S. C. 1991. A structural taxonomy of DNA binding domains. *Nature* 353, 715–19.

Hayashi, S. and M. P. Scott 1990. What determines the specificity of action of *Drosophila* homeodomain proteins. *Cell* 63, 883–94.

Hollenberg, S. M. and R. M. Evans 1988. Multiple and cooperative transactivation domains of the human glucocorticoid receptor. *Cell* 55, 899–906.

Hollenberg, S. M., V. Giguere, P. Segui and R. M. Evans 1987. Colocalization of DNA-binding and transcriptional activation functions in the human glucocorticoid receptor. *Cell* 49, 39–46.

Hope, I. A. and K. Struhl 1986. Functional dissection of a eukaryotic transcriptional activator GCN4 of yeast. *Cell* 46, 885–94.

Horikoshi, M., M. F. Carey, H. Kakidani and R. G. Roeder 1988. Mechanism of action of a yeast activator: direct effect of GAL4 derivatives on mammalian TFIID-promoter interactions. *Cell* 54, 665–9.

Hunter, T. and M. Karin 1992. The regulation of transcription by phosphorylation. *Cell* 70, 375–87.

Ingraham, P. W. 1988. The molecular genetics of embryonic pattern formation in *Drosophila*. *Nature* 335, 25–34.

Jackson, M.E. 1991. Negative regulation of eukaryotic transcription. *Journal of Cell Science* 100, 1–7.

Jackson, S. P. 1992. Regulating transcription factor activity by phosphorylation. *Trends in Cell Biology* 2, 104–8.

Johnson, P. F. and S. L. McKnight 1989. Eukaryotic transcriptional regulatory proteins. *Annual Review of Biochemistry* 58, 799–839.

Jones, N. 1990. Transcriptional regulation by dimerization: two sides to an incestuous relationship. *Cell* 61, 9–11.

Kadonaga, J. T. and R. Tjian 1986. Affinity purification of sequence-specific DNA binding proteins. *Proceedings of the National Academy of Sciences of the USA* 83, 5889–93.

Kadonaga, J. T., K. R. Carner, F. R. Masiarz and R. Tjian 1987. Isolation of cDNA encoding the transcription factor Spl and functional analysis of the DNA binding domain. *Cell* 51, 1079–90.

Karin, M. 1992. Signal transduction from cell surface to nucleus in development and disease. *FASEB Journal* 6, 2581–90.

Karin, M. and T. Smeal 1992. Control of transcription factors by signal transduction pathways: the beginning of the end. *Trends in Biochemical Sciences* 17, 418–22.

Kissinger, C. R., B. Liu, E. Martin-Blanco, T. B. Kornberg and C. O. Pabo 1990. Crystal structure of an engrailed homeodomain–DNA complex at 2.8 Å resolution: a framework for understanding homeodomain–DNA interactions. *Cell* 63, 579–90.

Klug, A. and D. Rhodes 1987. Zinc fingers: a novel protein motif for nucleic acid recognition. *Trends in Biochemical Sciences* 12, 464–9.

Kornberg, T. B. 1993. Understanding the homeodomain. *Journal of Biological Chemistry* 268, 26813–16.

Krasnow, M. A., E. C. Saffman, K. Kornfeld and D. S. Hogness 1989. Transcriptional activation and repression by ultrabithorax proteins in cultured Drosophila cells. *Cell* 57, 1031–43.

Lamb, P. and S. L. McKnight 1991. Diversity and specificity in transcriptional regulation: the benefits of heterotypic dimerization. *Trends in Biochemical Sciences* 16, 417–22.

Latchman, D. S. 1991. *Eukaryotic Transcription Factors*. Academic Press, London.

Lewin, B. 1990. Commitment and activation at pol II promoters: a tail of protein–protein interactions. *Cell* 61, 1161–4.

Lewin, B. 1991. Oncogenic conversion by regulatory changes in transcription factors. *Cell* **64**, 303–12.

Lillycrop, K. A. and D. S. Latchman 1992. Alternative splicing of the Oct-2 transcription factor is differentially regulated in neuronal cells and B cells and results in protein isoforms with opposite effects on the activity of octamer/TAATGARAT-containing promoters. *Journal of Biological Chemistry* **267**, 24960–6.

Miller, J., A. D. McLachlan and A. Klug 1985. Repetitive zinc-binding domains in the protein transcription factor III A from *Xenopus* oocytes. *EMBO Journal* **4**, 1609–14.

Mitchell, P. J. and R. Tjian 1989. Transcriptional regulation in mammalian cells by sequence specific DNA binding proteins. *Science* **245**, 371–8.

Muller, M. M., S. Ruppert, W. Schaffner and P. Matthias 1988. A cloned octamer transcription factor stimulates transcription from lymphoid-specific promoters in non-B cells. *Nature* **336**, 544–51.

Parker, M. G. 1993. Steroid and related receptors. *Current Opinion in Cell Biology* **5**, 499–504.

Ptashne, M. 1988. How eukaryotic transcriptional activators work. *Nature* **335**, 683–9.

Redemann, N., U. Gaul and H. Jackle 1988. Disruption of a putative Cys–zinc interaction eliminates the biological activity of the Kruppel finger protein. *Nature* **332**, 90–2.

Rhodes, D. and Klug, A. 1993. Zinc finger structure. *Scientific American* **268** (4), 32–39.

Ruiz-i-Altaba, A., H. Perry-O'Keefe and D. A. Melton 1987. Xfin: an embryonic gene encoding a multi-fingered protein in *Xenopus*. *EMBO Journal* **6**, 3065–70.

Schaffner, W. 1989. How do different transcription factors binding the same DNA sequence sort out their jobs? *Trends in Genetics* **5**, 37–9.

Schwabe, J. W. R. and D. Rhodes 1991. Beyond zinc fingers: steroid hormone receptors have a novel structural motif for DNA recognition. *Trends in Biochemical Sciences* **16**, 291–6.

Scott, M. P. and S. B. Carroll 1987. The segmentation and homeotic gene network in early *Drosophila* development. *Cell* **51**, 689–98.

Scott, M. P., J. W. Tamkun and G. W. Hartzell 1989. Structure and function of the homeodomain. *Biochimica et Biophysica Acta* **989**, 25–48.

Sharp, P. A. 1991. TFIIB or not TFIIB? *Nature* **351**, 16–18.

Shepherd, J. C. W., W. McGinnis, A. E. Carrasco, E. M. De Robertis and W. J. Gehring 1984. Fly and frog homeo domains show homologies with yeast mating type regulatory loci. *Nature* **310**, 70–1.

Sigler, P. B. 1988. Acid blobs and negative noodles. *Nature* **333**, 210–12.

Singh, H., J. H. Le Bowitz, A. S. Baldwin and P. A. Sharp 1988. Molecular cloning of an enhancer binding protein: isolation by screening of an expression library with a recognition site DNA. *Cell* **52**, 415–29.

Struhl, K. 1989. Helix-turn-helix, zinc finger and leucine zipper motifs for eukaryotic transcriptional regulatory proteins. *Trends in Biochemical Sciences* **14**, 137–40.

Tjian, R. and T. Maniatis 1994. Transcriptional activation: a complex puzzle with few easy pieces. *Cell* 77, 5–8.

Travers, A. 1993. *DNA–Protein Interactions*. Chapman & Hall, London.

Umesono, K. and R. M. Evans 1989. Determinants of target gene specificity for steroid/thyroid hormone receptors. *Cell* 57, 1139–46.

Verrijzer, C.P and P. C. van der Vliet 1993. POU domain transcription factors. *Biochimica et Biophysica Acta* **1173**, 1–21.

Wegner, M., D. W. Drolet and M. G. Rosenfeld 1993. POU-domain proteins: structure and function of developmental regulators. *Current Opinion in Cell Biology* **5**, 488–98.

Xanthopoulos, K. G., J. Mirkovitch, T. Decker, C. F. Kuo and J. E. Darnell, Jr 1989. Cell-specific transcriptional control of the mouse DNA binding protein mC/EBP. *Proceedings of the National Academy of Sciences of the USA* **86**, 4117–21.

Young, R. A. and R. W. Davis 1983. Yeast RNA polymerase II genes: isolation with antibody probes. *Science* **222**, 778–82.

Gene regulation and cancer

8

8.1 Introduction

As we have discussed in the preceding chapters, the regulation of gene expression in higher eukaryotes is a highly complex process. It is not surprising, therefore, that this process can go wrong, and the identification of the molecular basis of many human diseases has shown some to be due to defects in gene regulation. For example, one form of congenital severe combined immunodeficiency is characterized by a lack of expression of HLA class II genes and has been shown to be due to the lack of a specific transcription factor necessary for the transcription of these genes (Reith *et al.*, 1988). Similarly, mutations in the gene encoding the Pit-l member of the POU family of transcription factors (Section 7.2.2) result in a failure of pituitary gland development and consequent dwarfism in both mice and humans (Radovick *et al.*, 1992).

As well as affecting transcription factors, mutations can also result in disease if they affect other regulatory processes, either transcriptional or post-transcriptional. Thus in the human fragile X syndrome a triplet sequence CGG which is present in 10–50 tandem copies in the first exon of the FMR-l gene is amplified so that over 230 tandem copies are present in patients with this syndrome. This amplified sequence is then heavily methylated on the C residues (Section 5.5.1), resulting in transcriptional silencing of the FMR-l gene which in turn produces the mental retardation and other symptoms characteristic of the fragile X syndrome (for review, see Trottier *et al.*, 1993). Similarly, the failure to produce one of the two alternatively spliced mRNAs derived from the porphobillinogen deaminase gene is the cause of one form of the disease acute intermittent porphyria (Grandchamp *et al.*,1989).

The human disease that exhibits the most extensive malregulation of gene expression, however, is cancer. Thus, not only does cancer often result from the overexpression of certain cellular genes (known as proto-oncogenes) due to errors in their regulation, but several of these genes themselves actually encode transcription factors and cause the disease by affecting the expression of other genes. This chapter will therefore focus on the connection between cancer and the malregulation of gene expression, and will illustrate how our increasing knowledge of this connection has aided our understanding both of the disease and of the processes that regulate gene expression in normal and transformed cancer cells.

8.2 Proto-oncogenes

In order to provide a background to our discussion of cancer and gene regulation, it is necessary to briefly discuss the nature of cancer-causing oncogenes and the processes that led to their discovery (reviewed by Weinberg, 1985; Bishop, 1987).

As long ago as 1911, Peyton Rous showed that a connective tissue cancer in the chicken was caused by an infectious agent. This agent was subsequently shown to be a virus and was named Rous Sarcoma virus (RSV) after its discoverer. This cancer-causing, tumorigenic virus is a member of a class of viruses called retroviruses, whose genome consists of RNA rather than DNA as in most other organisms.

The majority of viruses of this type do not cause cancers but simply infect a cell and produce a persistent infection with continual production of virus by the infected cell. In the case of RSV, however, such infection also results in the conversion of the cell into a cancer cell, capable of indefinite growth and eventually killing the organism containing it.

In the case of non-tumorigenic retroviruses, the genome contains only three genes, which are known as *gag*, *pol* and *env*, and which function in the normal life cycle of the virus (Figure 8.1). Following cellular entry, the viral RNA is converted into DNA by the action of the Pol protein and this DNA molecule then integrates into the host chromosome. Subsequent transcription and translation of this DNA produces the viral structural proteins Gag and Env, which coat the viral RNA genome, yielding viral particles that leave the cell to infect other susceptible cells.

Inspection of the RSV genome (Figure 8.2) reveals an additional gene, known as the *src* gene, which is absent in the other viruses, suggesting that this gene is responsible for the ability of the virus to cause cancer. This idea was confirmed subsequently by showing that if the *src* gene alone was introduced into normal cells it was able to transform them to a cancerous phenotype. This gene was therefore called an oncogene (from the Greek *onkas* for mass or tumor) or cancer-causing gene.

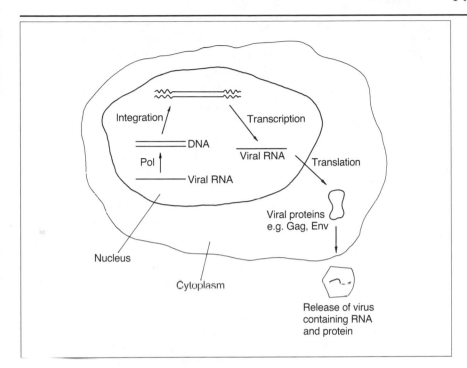

Figure 8.1 Life cycle of a typical non-tumorigenic retrovirus.

Following the identification of the *src* oncogene in RSV, a number of other oncogenes were identified in other oncogenic retroviruses infecting both chickens and mammals, such as the mouse or rat. Over 20 cellular proto-oncogenes which were originally identified in this way are now known (Table 8.1).

The identification of individual genes that are able to cause cancer obviously opened up many avenues of investigation for the study of this disease. From the point of view of gene regulation, however, the most exciting aspect of oncogenes was provided by the discovery that these cancer-causing genes are derived from genes present in normal cellular DNA. Thus, using Southern blotting techniques (Section 2.2.4), Takeya and Hanafusa (1985) detected a cellular equivalent of the viral *src* gene in

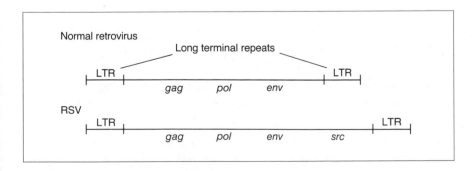

Figure 8.2 Comparison of the genome of a non-tumorigenic retrovirus with that of the tumorigenic retrovirus RSV.

Table 8.1 Proto-oncogenes and their functions

Oncogene	Species infected by virus	Normal function of protein
abl	Mouse	Tyrosine kinase
erbA	Chicken	Transcription factor and hormone receptor
erbB	Chicken	Receptor for epidermal growth factor, tyrosine kinase
ets	Chicken	Transcription factor
fes	Cat	Tyrosine kinase
fgr	Cat	Tyrosine kinase
fms	Cat	Tyrosine kinase, receptor for colony-stimulating factor
fos	Mouse	Transcription factor
jun	Chicken	AP1-related transcription factor
kit	Cat	Tyrosine kinase
lck	Chicken	Tyrosine kinase
mos	Mouse	Serine/threonine kinase
myb	Chicken	Transcription factor
myc	Chicken	Transcription factor
raf	Chicken	Serine/threonine kinase
ras	Rat	GTP-binding protein
rel	Turkey	Nuclear protein
ros	Chicken	Tyrosine kinase
sea	Chicken	Tyrosine kinase
sis	Monkey	Platelet-derived growth factor B chain
ski	Chicken	Nuclear protein
src	Chicken	Tyrosine kinase
yes	Chicken	Tyrosine kinase

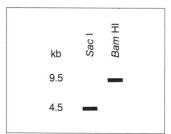

Figure 8.3 Southern blot showing hybridization of the RSV *src* gene with normal chicken DNA cut with the restriction enzymes *Sac*I or *Bam*HI.

the DNA of both normal and cancer cells (Figure 8.3) and also showed that an mRNA capable of encoding the Src protein was produced in normal cells. Subsequent studies identified cellular equivalents of all the retroviral oncogenes. Many of these cellular equivalents of the viral oncogenes have now been cloned and shown to encode proteins identical or closely related to those present in the retroviruses. To avoid confusion, the viral oncogenes are given the prefix v, as in v-*src*, while their cellular equivalents are designated proto-oncogenes and given the prefix c, as in c-*src*.

The presence of cellular equivalents of the retroviral oncogenes suggests that the viral genes have been picked up from normal cellular DNA following integration of a non-tumorigenic virus next to the proto-oncogene. The subsequent incorrect excision of the viral genome resulted in the virus picking up the proto-oncogene and rendered the virus tumorigenic

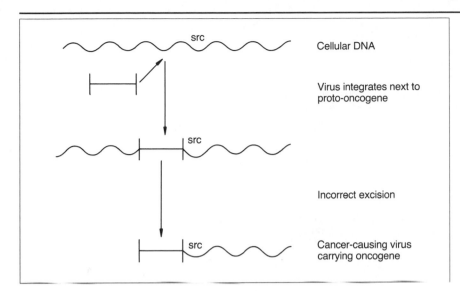

Figure 8.4 Model for the way in which a retrovirus could have picked up the cellular *src* gene by integration adjacent to the gene and subsequent incorrect excision.

(Figure 8.4). Hence the oncogene in the virus is derived from a cellular gene which is present in the DNA of normal cells.

Paradoxically the same gene can be present in normal cells without any apparent adverse effect and yet can cause cancer when incorporated into a virus. It is now clear that this conversion of the proto-oncogene into a cancer-causing oncogene is caused by one of two processes. Either the gene is mutated in some way within the virus, so that an abnormal product is formed, or alternatively the gene is expressed within the virus at much higher levels than are achieved in normal cells and this high level of the normal product results in transformation (Figure 8.5).

Such conversion of a proto-oncogene into a cancer-causing oncogene is not confined to viruses, however. Cases of cancer which have no viral involvement in humans have now been shown to be due either to the overexpression of individual cellular proto-oncogenes or to their alteration by mutation within the cellular genome. Hence these genes play an important role in the generation of human cancer.

The potential risk of such proto-oncogenes causing cancer raises the question of why these genes have not been deleted during evolution. In fact, proto-oncogenes have been highly conserved in evolution, equivalents to mammalian and chicken oncogenes having been found not only in other vertebrates but also in invertebrates, such as *Drosophila*, and even in single-celled organisms, such as yeast (see, e.g. De Feo-Jones *et al.*, 1983).

The extraordinary evolutionary conservation of many of these genes, despite their potential danger, led to the suggestion that their products were essential for the processes regulating the growth of normal cells and

Figure 8.5 A cellular proto-oncogene can be converted into a cancer-causing oncogene by increased expression or by mutation.

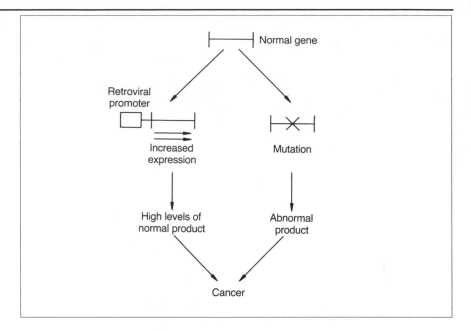

that their malregulation or mutation therefore results in abnormal growth and cancer. This idea has been confirmed abundantly as more and more proto-oncogenes have been characterized and shown to encode growth factors that stimulate the growth of normal cells, cellular receptors for growth factors and other cellular proteins involved in transmitting the growth signal within the cell, either by acting as a protein kinase enzyme or by binding GTP (for a review, see Bourne and Varmus, 1992).

Ultimately the growth regulatory pathways controlled by oncogene products end in the nucleus, with the activation of genes whose corresponding proteins are required by the growing cell. It is not surprising, therefore, that several proto-oncogenes have been shown to encode transcription factors that regulate the expression of genes activated in growing cells (for reviews, see Lewin, 1991; Forrest and Curran, 1992) (Table 8.2).

Hence oncogenes present two aspects of importance from the point of view of the regulation of gene expression. First, since cancer is often caused by elevated expression of cellular oncogenes, the processes whereby this occurs are of interest, both from the point of view of the etiology of cancer and for the light they throw on the mechanisms which regulate gene expression. This topic is therefore discussed in Section 8.3. Secondly, the study of the transcription factors encoded by a few proto-oncogenes has led to a better understanding of the processes regulating gene expression in both cells growing normally and in cancer cells, and this is discussed in Section 8.4.

Table 8.2 Oncogenes and transcription factors

Oncogene	Comments
erbA	Mutant form of the thyroid hormone receptor
ets	Binding site often found in association with AP1 site
fos	Binds to AP1 site as Fos–Jun dimer
jun	Can bind to AP1 site alone as Jun–Jun dimer
mdm2	Inhibits gene activation by p53
myb	–
myc	Requires Max protein to bind to DNA
rel	Member of the NF κB family
spi-1	Identical to PU.1 transcription factor

8.3 Elevated expression of oncogenes

The products of cellular proto-oncogenes play a critical role in cellular growth control and, in many cases, are synthesized only at specific times and in small amounts. It is not surprising, therefore, that transformation into a cancer cell can result when these genes are expressed at high levels in particular situations.

The simplest example of such overexpression occurs in the case of retroviruses where, as we have already discussed, the oncogene comes under the influence of the strong promoter contained in the retroviral long terminal repeat (LTR) region and is hence expressed at a high level (Figure 8.5).

A similar up-regulation due to the activity of a retroviral promoter is also seen in the case of avian leukosis virus (ALV) of chickens. Unlike the retroviruses described so far, however, this virus does not carry its own oncogene. Rather it transforms by integrating into cellular DNA next to a cellular oncogene, the c-myc gene (Figure 8.6a) (for review of c-myc, see Spencer and Groudine, 1991). The expression of the c-myc gene is brought under the control of the strong promoter in the retroviral LTR and it is hence expressed at levels 20–50 times higher than normal, producing transformation (Hayward et al., 1981). This process is known as promoter insertion.

In other cases of this type, the ALV virus has been shown to have integrated downstream rather than upstream of the c-myc gene. Hence the elevated expression of the c-myc gene in these cases cannot be due to promoter insertion. Rather, it involves the action of an enhancer element in the viral LTR which activates the c-myc gene's own promoter. As discussed in Chapter 6 (Section 6.3.1), enhancers, unlike promoters, can act in either orientation and at a distance, and hence the ALV enhancer

Figure 8.6 Avian leukosis virus (ALV) can increase expression of the *myc* proto-oncogene either via promoter insertion (a) or by the action of its enhancer (b).

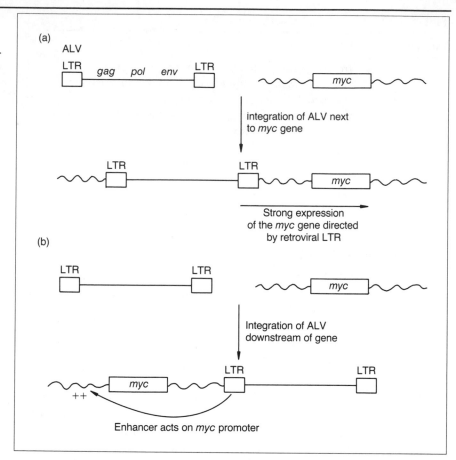

can activate the *myc* promoter readily from a position downstream of the gene (Figure 8.6b).

These cases are of interest from the point of view of gene regulation, as indicating how viral regulatory systems can subvert cellular control processes. Of potentially greater interest, however, are the cases where up-regulation of a cellular oncogene occurs through the alteration of internal cellular regulatory processes, rather than through viral intervention. The best-studied example of this type concerns the increased expression of the c-*myc* oncogene which occurs in the transformation of B cells in cases of Burkitt's lymphoma in humans or in the similar plasmacytomas that occur in mice.

When these tumors were studied, it was noted that they commonly contained very specific chromosomal translocations which involved the exchange of genetic material between chromosome 8 and chromosome 14 (Figure 8.7). Most interestingly, the region of chromosome 8 involved includes the c-*myc* gene and the translocation results in the *myc* gene being

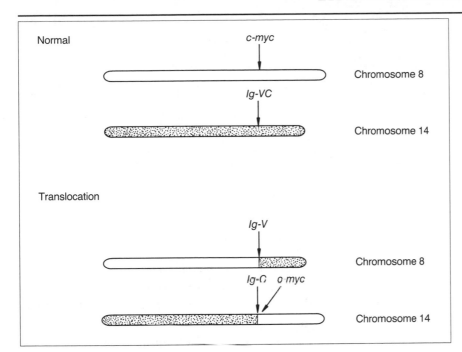

Figure 8.7 Translocation of the c-*myc* gene from chromosome 8 to the immunoglobulin heavy-chain gene locus on chromosome 14 that occurs in cases of Burkitt's lymphoma.

moved to chromosome 14, where it becomes located adjacent to the gene encoding the immunoglobulin heavy chain. This translocation results in the increased expression of the *myc* gene which is observed in the tumor cells.

Detailed study of the processes mediating such increased expression has indicated that it is produced by different mechanisms in different lymphomas, depending on the precise break points of the translocation within the c-*myc* and immunoglobulin genes. In all cases studied, however, the break point of the translocation occurs within the immunoglobulin gene, resulting in a truncated gene lacking its promoter being linked to the c-*myc* gene. This fact, together with the fact that the genes are always linked in a head-to-head orientation (Figure 8.8), indicates that the up-regulation of the c-*myc* gene does not occur via a simple promoter

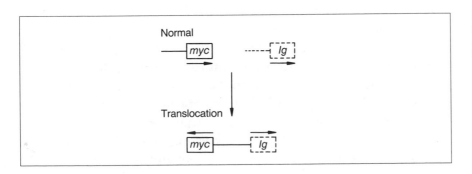

Figure 8.8 Head-to-head orientation of the translocated *myc* gene and the immunoglobulin gene.

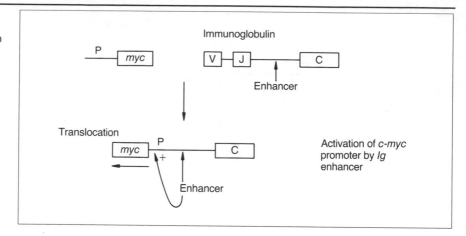

insertion mechanism (see Figure 8.6a) in which it comes under the control of the immunoglobulin promoter. In some cases, however, the B-cell-specific enhancer element, which is located between the joining and constant regions of the immunoglobulin genes (see Section 2.4 and Section 6.3.2), is brought close to the *myc* promoter by the translocation (Figure 8.9). This enhancer element is highly active in B cells and can activate the myc promoter in a manner analogous to the enhancer of ALV (see Figure 8.6b).

Hence in this case the c-*myc* gene is up-regulated by the action of the B-cell-specific regulatory mechanisms of the immunoglobulin gene, the immunoglobulin enhancer activating the c-*myc* promoter rather than its own promoter, which has been removed by the translocation. In other cases, however, this does not appear to be the case, the break point of the translocation having removed both the immunoglobulin promoter and enhancer. This leaves the c-*myc* gene adjacent to the constant region of the immunoglobulin gene without any obvious B-cell-specific regulatory elements. In these cases it is likely that the up-regulation of the c-*myc* gene arises from its own truncation in the translocation. This results in the removal of negative regulatory elements that normally repress its expression, such as the upstream silencer element which normally represses the c-*myc* promoter (see Section 6.3.4 and Figure 8.10).

More extensive truncation of the c-*myc* gene, involving the removal of transcribed sequences rather than upstream elements, has also been observed in some tumors. Frequently, this involves the removal of the first exon of the c-*myc* gene which does not contain any protein-coding information. This exon may thus fulfil a regulatory role by modulating the stability of the c-*myc* RNA or by affecting its translatability. Hence its removal could enhance the level of c-*myc* protein by increasing the

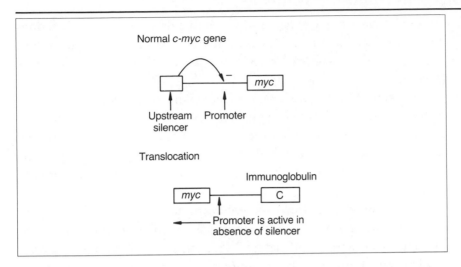

Figure 8.10 Activation of *myc* gene expression can be achieved by removal of the upstream silencer element.

stability or the efficiency of translation of the c-*myc* RNA produced at a constant level of transcription. A similar increase in gene expression could be achieved by the removal of sequences within the first intervening sequence which inhibit the transcriptional elongation of the nascent c-*myc* transcript (Section 3.3.3).

The increased expression of an oncogene produced by the removal of sequences that negatively regulate it is also seen in the case of the *lck* proto-oncogene, which encodes a tyrosine kinase related to the c-*src* gene product. In this case, activation of the oncogene in tumors is accompanied by the removal of sequences within its 5' untranslated region, upstream of the start site of translation (Marth *et al.*, 1988). The removal of these sequences results in a 50-fold increase in the initiation of translation of the *lck* mRNA into protein. Most interestingly, the region removed contains three AUG translation initiation codons which are located upstream of the correct initiation codon for production of the Lck protein (Figure 8.11). The elimination of these codons results in increased translation initiation

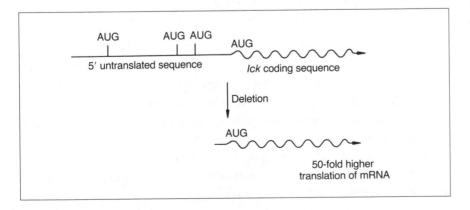

Figure 8.11 Increased translation of the *lck* proto-oncogene mRNA can be achieved by deletion of AUG translational start sites upstream of the AUG that initiates the coding sequence of the Lck protein.

from the correct AUG, suggesting that initiation at the upstream codons inhibits correct initiation. This is exactly analogous to the regulation of translation of the GCN4 protein, which was discussed in Chapter 4 (Section 4.5.2).

It is likely that the processes regulating the expression of oncogenes such as c-*myc* and *lck*, which are revealed by studying their overexpression in tumors, also play a role in their normal pattern of regulation during cellular growth. Hence their further study will throw light not only on the mechanisms of tumorigenesis but also on the processes regulating gene expression in normal cells.

Other examples of up-regulation of oncogene expression in tumorigenesis may occur, however, by mechanisms that are unique to the transformed cell. Thus, as discussed in Chapter 2 (Section 2.3), DNA amplification is relatively rare in normal cells. In tumors, however, it is observed frequently for specific oncogenes (Collins and Groudine, 1982) and results in the presence of regions of amplified DNA which are visible in the microscope as homogeneously staining regions or as double minute chromosomes. Such amplification is especially common in human lung tumors and brain tumors, and frequently involves the c-*myc*-related genes, N-*myc* and L-*myc*. The expression of these genes in the tumor is increased dramatically due to the presence of up to 1000 copies of the gene in the tumor cell.

A variety of mechanisms, involving both the subversion of normal control processes or abnormal events occurring in the tumor cell, therefore result in the observed overexpression of oncogenes in tumor cells. When taken together with the production of abnormal oncogene products due to mutations, it is likely that these processes are involved in the majority of human cancers.

8.4 Transcription factors as oncogenes

As described in Section 8.2, the isolation of the cellular genes encoding particular oncogene products led to the realization that they were involved in many of the processes regulating cellular growth. Ultimately the onset and continuation of cellular growth is likely to involve the activation of cellular genes that are not expressed in quiescent cells. It is not surprising, therefore, that several proto-oncogenes have been shown to encode transcription factors which regulate the transcription of genes activated in growing cells (Table 8.2) and several of these cases will be discussed (for reviews, see Lewin, 1991; Forrest and Curran, 1992).

```
jun    206 P L F P I D M E S Q E R I K A E R K R M R N R I A A S K S R K
GCN4 216 P L S P I V P E S S D P   A A L K R A R N T E A A R R S R A

          R K L E R I A R L E E K V K T L K A Q N S E L A S T A N M L R
          R K L Q R M K Q L E D K V             E E L L S K N Y H L E

          E Q V A Q L K Q K V M N H V N S G C Q L M L T Q Q L G T F 296
          N E V A R L K K L V G E R 281
```

Figure 8.12 Comparison of the C-terminal amino acid sequences of the chicken Jun protein and the yeast transcription factor GCN4. Boxes indicate identical residues.

8.4.1 Fos, Jun and AP1

The chicken retrovirus, avian sarcoma virus ASV17, contains an oncogene, v-*jun*, whose equivalent cellular proto-oncogene encodes a nuclearly located DNA-binding protein. Sequence analysis of this protein revealed that it showed significant homology to the DNA-binding domain of the yeast transcription factor GCN4, suggesting that it might bind to similar DNA sequences (Figure 8.12; Vogt *et al.*, 1987). Interestingly, GCN4 itself had been shown previously to bind to similar sequences to those bound by a factor, AP1 (activator protein 1), which had been detected in mammalian cell extracts by its DNA-binding activity (Figure 8.13).

This relationship of Jun and AP1 to the sequence and binding activity, respectively, of GCN4 led to the suggestion that Jun might be related to AP1. This was confirmed by the findings that antibody to Jun reacted with purified AP1 preparations and that Jun expressed in bacteria was capable of binding to AP1-binding sites in DNA (Bos *et al.*, 1988). Moreover, Jun was capable of stimulating transcription from promoters containing AP1-binding sites but not from those which lacked these sites (Angel *et al.*, 1988). Hence the *jun* oncogene encodes a sequence-specific DNA-binding protein capable of stimulating transcription of genes containing its binding site, which is identical to the AP1-binding site.

Although Jun undoubtedly binds to AP1-binding sites, preparations of AP1 purified on the basis of this ability contain several other proteins in addition to c-Jun. Several of these are encoded by genes related to *jun*, but another is the product of a different proto-oncogene, i.e. c-*fos*. Interestingly, however, although Fos is present in AP1 preparations, it does not

```
                    DNA binding site

          GCN4  5' T G A C/G T C A T 3'
           AP1  3' T G A G   T C A G 3'
```

Figure 8.13 Relationship of the DNA-binding sites for the yeast transcription factor GCN4 and the mammalian transcription factor AP1.

bind to DNA when present alone but requires the product of the c-*jun* gene for DNA binding (Rauscher *et al.*,1988). Hence, in addition to its ability to bind to APl sites alone, Jun can also form a complex with Fos that binds to this site. As discussed in Chapter 7 (Section 7.2.4), this association takes place through the leucine zipper domains of the two proteins and results in a heterodimer complex which binds to the APl-binding site with much greater affinity than the Jun homodimer (Hala-zonetis *et al.*,1988).

Hence both Fos and Jun, which were identified originally through their association with oncogenic retroviruses, are also cellular transcription factors (for reviews, see Curran and Franza, 1988; Ransone and Verma, 1990). Such a finding raises the question of the normal role of these factors and how they can cause cancer. In this regard it is of obvious inter-est that the APl-binding site is involved in mediating the induction of genes that contain it in response to treatment with phorbol esters, which are also capable of promoting cancer. Thus not only do many phorbol ester-inducible genes contain APl-binding sites (see Table 6.3) but, in addition, transfer of APl-binding sites to a normally non-inducible gene renders that gene inducible by phorbol esters (Angel *et al.*, 1987; Lee *et al.*, 1987). Increased levels of Jun and Fos are also observed in cells after treat-ment with phorbol esters (Lamph *et al.*, 1988), indicating that these substances act by increasing the levels of Fos and Jun, which in turn cause increased transcription of other genes containing APl-binding sites which mediate induction by the Fos–Jun complex.

Most interestingly, increased levels of Jun and Fos are also produced by treatment with serum or growth factors which stimulates the growth of quiescent cells. Hence the transduction of the signal to grow, which begins with the growth factors and their cellular receptors and continues with intra-cytoplasmic signal transducers such as protein kinases and GTP-binding proteins, ends in the nucleus with the increased level of the tran-scription factors Jun and Fos (Figure 8.14). These proteins will then activate the genes whose products are necessary for the process of growth itself.

Clearly, it is relatively easy to fit the oncogenic properties of Jun and Fos into this framework. Thus, if these proteins are normally produced in response to growth-inducing signals and activate growth, their continual synthesis will result in a cell which will be stimulated to grow continually and will not respond to growth-regulating signals. Such continuous uncontrolled growth is characteristic of the cancer cell.

In agreement with this idea, mutations in the leucine zipper region of Fos, which abolish its ability to dimerize with Jun and induce genes

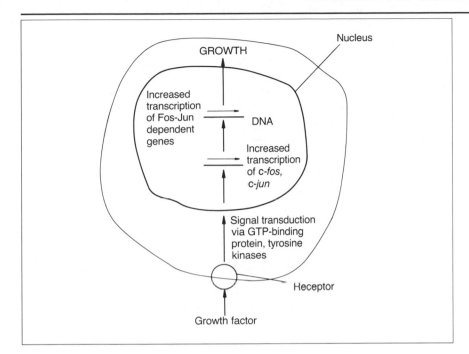

Figure 8.14 Growth factor stimulation of cells results in increased transcription of the c-*fos* and c-*jun* genes, which in turn stimulates transcription of genes which are activated by the Fos–Jun complex.

containing AP1 sites, also abolish its ability to transform cells to a cancerous phenotype (Schuermann *et al.*, 1989). Hence, the ability of Fos to cause cancer is directly linked to its ability to act as a transcription factor for genes containing the appropriate binding site.

Interestingly, AP1 sites in growth-regulated genes are often located close to binding sites for the Ets protein, another transcription factor which is encoded by a cellular proto-oncogene (see Table 8.2). Hence several different oncogenic transcription factors may co-operate to produce high-level transcription of specific genes in actively growing cells. In contrast, gene activation by the Fos–Jun complex can be inhibited by the presence of activated glucocorticoid receptor which has dissociated from the inhibitory hsp90 protein in the presence of glucocorticoid hormone (Section 7.4.3). In this case the Fos–Jun complex interacts with the glucocorticoid receptor by protein–protein interaction. This results in a failure of both proteins to bind to DNA and hence prevents activation of both Fos–Jun-dependent genes and the glucocorticoid-dependent genes (Figure 8.15). This mutual repression therefore illustrates the manner in which transcription factors activated by different signaling pathways can interact with one another (for review, see Schule and Evans, 1991). In addition it illustrates how the same transcription factor can act either positively or negatively depending on the situation. Thus both AP1 and the activated glucocorticoid receptor can activate transcription from genes containing

Figure 8.15 Interaction of the API (Fos–Jun) factor with the glucocorticoid receptor to form a protein–protein complex which does not bind to DNA and hence cannot activate transcription via either an API-binding site or a glucocorticoid response element.

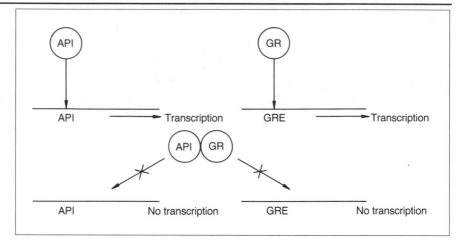

their binding sites when present alone but each factor is also able to inhibit activation by the other factor via protein–protein interactions which prevent DNA binding when both are present together (see Section 7.3.4).

8.4.2 v-*erbA* and the thyroid hormone receptor

Unlike most other retroviruses, avian erythroblastosis virus (AEV) carries two cellular oncogenes, v-*erbA* and v-*erbB* (reviewed by Beug *et al.*, 1985). When c-*erbA*, the cellular equivalent of v-*erbA*, was cloned, it was shown to encode the cellular receptor that mediates the response to thyroid hormone (Sap *et al.*, l986; Weinberger *et al.*, 1986).

As discussed in Chapter 7 (Sections 7.3.4 and 7.4.2), this receptor is a member of the super-family of steroid–thyroid hormone receptors which, following binding of a particular hormone, induce the transcription of genes containing a binding site for the hormone–receptor complex. In the case of ErbA, the protein contains a region that can bind thyroid hormone. Following such binding, the hormone–receptor complex induces transcription of thyroid-hormone-responsive genes, such as those encoding growth hormone or the heavy chain of the myosin molecule, by binding to its appropriate DNA recognition site within their promoters (Figure 8.16).

The finding that the cellular homologue of the v-*erbA* oncogene is a hormone-responsive transcription factor therefore provides a further connection between oncogenes and cellular transcription factors. It raises the question, however, of the manner in which the transfer of the thyroid hormone receptor to a virus can result in transformation. To answer this question it is necessary to compare the protein encoded by the virus with its cellular counterpart. As shown in Figure 8.17, the ErbA protein has the typical structure of a member of the steroid–thyroid hormone family (see

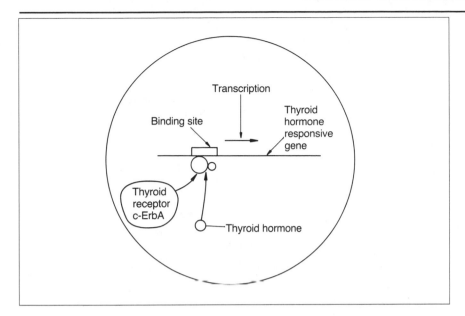

Figure 8.16 The c-*erbA* gene encodes the thyroid hormone receptor and activates transcription in response to thyroid hormone.

also Figure 7.14), containing both DNA-binding and hormone-binding regions. The viral ErbA protein is generally similar except that it is fused to a portion of the retroviral gag protein at its N-terminus. It also contains a number of mutations in both the DNA-binding and hormone-binding regions, as well as a small deletion in the hormone-binding domain.

Of these changes, it is the alterations in the hormone-binding domain which have the most significant effects on the function of the protein and which are thought to be critical for transformation. Thus these changes abolish the ability of the protein to bind thyroid hormone and activate transcription. However, the protein retains the inhibitory domain which allows the ErbA protein to repress gene transcription in the absence of thyroid hormone (see Section 7.3.4). Thus the viral protein is functionally analogous to the alternatively spliced form of the c-*erbA* gene product discussed in Chapter 7 (Section 7.4.2), which lacks the hormone-binding

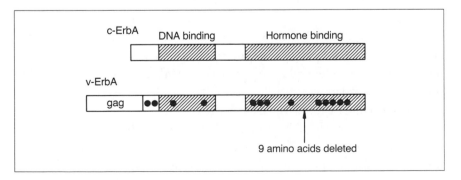

Figure 8.17 Relationship of the cellular ErbA protein and the viral protein. The black dots indicate single amino acid differences between the two proteins while the arrow indicates the region where nine amino acids are deleted in the viral protein.

Figure 8.18 Inhibitory effect of the viral ErbA protein on gene activation by the cellular protein, in response to thyroid hormone. Note the similarity to the action of the α–2 form of the c-ErbA protein, illustrated in Figure 7.37.

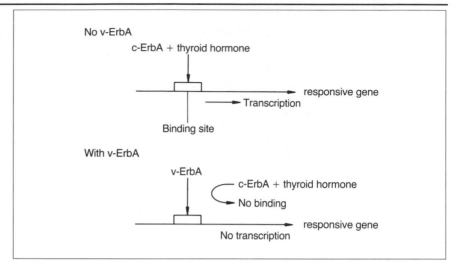

domain and dominantly represses the ability of the hormone-binding receptor form to activate thyroid-hormone-responsive genes.

The idea that the non-hormone-binding viral ErbA protein might also be able to do this has been confirmed by studying the effect of this oncogene on thyroid-hormone-responsive genes (Sap *et al.*, 1989). As expected, the v-*erbA* gene product was able to abolish the responsiveness of such genes to thyroid hormone by binding to the thyroid-hormone-response elements in their promoters and preventing binding of the cellular ErbA protein–thyroid hormone complex (Figure 8.18).

The explanation of how such gene repression by viral ErbA can result in transformation is provided by the observations of Zenke *et al.* (1988) who showed that the introduction of the viral gene into cells can repress transcription of the avian erythrocyte anion transporter gene. This gene is one of those which is switched on when chicken erythroblasts differentiate into erythrocytes. It has been known for some time that the viral ErbA protein can block this process, and it is now clear that this is achieved by blocking the induction of the genes needed for this to occur. In turn, such blockage of differentiation allows the cells to continue to proliferate. When this is combined with the introduction of the v-*erbB* gene, which encodes a truncated form of the epidermal growth factor receptor and renders cell growth independent of external growth factors, transformation results (Figure 8.19).

Transformation caused by v-*erbA* thus represents an example of the activation of an oncogene by mutation, resulting, in this case, in its ability to act as a dominant repressor of transcription. As discussed in Chapter 7 (Section 7.4.2), however, one alternatively spliced transcript of the c-*erbA*

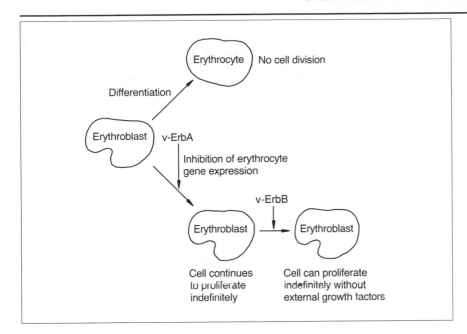

Figure 8.19 Inhibition of erythrocyte-specific gene expression by the v-ErbA protein prevents erythrocyte differentiation and allows transformation by the v-ErbB protein.

gene is also able to do this and, like the *v-erbA* gene product, cannot bind thyroid hormone. Hence this repression of transcription by a non-hormone-binding form of the receptor is likely to be of importance in normal cells also.

Indeed, such repression may be one facet of a complex network of regulatory controls involving the thyroid hormone receptor in intact cells. Thus the DNA sequence element mediating response to retinoic acid is identical to that which renders a gene inducible by thyroid hormone (Table 6.3). In agreement with this, retinoic acid bound to its cellular receptor can activate transcription from thyroid-hormone-responsive genes. In the absence of thyroid hormone, however, the thyroid hormone receptor binds to this site and blocks the binding of the retinoic acid complexed with its receptor (Graupner *et al.*, 1989). Since the thyroid hormone receptor in the absence of thyroid hormone acts as a dominant repressor of transcription, this has the effect of preventing transcription (Figure 8.20). This inhibition of retinoic acid-induced transcription by an unoccupied thyroid hormone receptor is clearly similar to the blockage of thyroid-hormone-induced transcription by a form of the receptor that cannot bind hormone. Hence the dominant repression of transcription produced by the *v-erbA* gene product has opened up a new area involving the interaction of different receptors and different forms of the same receptor in the regulation of gene expression.

Figure 8.20 The thyroid hormone receptor can bind to its DNA binding site in the absence of thyroid hormone and prevent binding of the retinoic acid–retinoic acid receptor complex to the identical binding site, thereby preventing gene activation in response to retinoic acid.

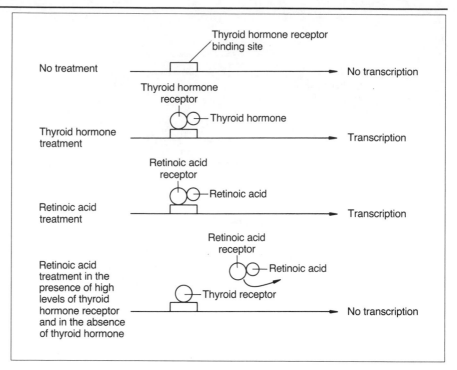

8.4.3 Other transcription-factor-related oncogenes

Although the *fos–Jun* and *erbA* cases represent the best examples of the connection between oncogenes and transcription factors, several other cellular oncogenes encode nuclear proteins which are likely to be transcription factors (see Table 8.2; for review, see Lewin, 1991).

One of these, the Myc protein, has been studied intensively in view of its overexpression in many human tumors (Section 8.3) (for review, see Spencer and Groudine, 1991). The Myc protein clearly encodes a transcription factor which contains the leucine zipper motif characteristic of many transcription factors, including Fos and Jun (see Section 7.2.4), as well as a helix-loop-helix motif. Moreover, mutations in the leucine zipper region of the protein abolish its oncogenic ability to transform normal cells, suggesting that the ability to act as a transcription factor is essential for Myc-induced transformation.

Despite all this evidence, for many years the actual role of Myc in transcriptional control remained unclear. This was because it was not possible to demonstrate the binding of Myc to a specific DNA sequence in the manner that had been shown to occur for Jun and ErbA. This problem was resolved, however, by the finding that Myc has to heterodimerize with a second factor Max in order to bind to DNA (for review, see Blackwood *et al.*, 1992). Hence Myc resembles Fos in requiring another factor for sequence-specific binding. This finding indicates once again the importance of heterodimerization in regulating the activity of transcription factors and also illustrates how the function of a particular factor can remain obscure simply because its partner has not yet been isolated.

Although our discussion of oncogenes as transcription factors has focused on the cellular genes which have been picked up by RNA tumor viruses, it is worth noting that DNA viruses that can cause cancer also encode oncogenes capable of regulating cellular gene expression (reviewed by Moran, 1993). In particular, both the large T oncogenes of the small DNA viruses, SV40 and polyoma, and the Ela protein of adenovirus are capable of affecting the transcription of specific cellular genes, and this ability is critical for the ability of these proteins to transform cells. Unlike the oncogenes of RNA viruses, however, the genes encoding these viral proteins do not appear to have specific cellular equivalents and are likely to have evolved within the virus rather than having been picked up from the cellular genome.

The similar ability of oncogenes from DNA and RNA tumor viruses to affect cellular gene expression, despite their very different origins, suggests that such modulation of cellular gene expression is critical for the transforming ability of these viruses.

8.5 Anti-oncogenes

8.5.1 Nature of anti-oncogenes

Following the discovery of cellular oncogenes it was rapidly shown that they encoded proteins which promoted cellular growth so that their activation by overexpression or by mutation resulted in abnormal growth leading to cancer. Subsequently, however, it became clear that cancer could also result from the deletion or mutational inactivation of another group of genes. This indicated that these genes encoded products which normally restrained cellular growth so that their inactivation would result in abnormal, unregulated growth. These genes were therefore named anti-oncogenes or tumor suppressor genes (for reviews, see Knudson, 1993; Weinberg, 1993).

Clearly the inhibitory role of anti-oncogenes indicates that cancers will involve the deletion of these genes or the occurrence of mutations within them which inactivate their protein product, rather than their overexpression or mutational activation as occurs in the oncogenes (Figure 8.21). Thus far a number of genes of this type have been defined on the basis

Figure 8.21 Deletion of an anti-oncogene or its inactivation by mutation can result in cancer.

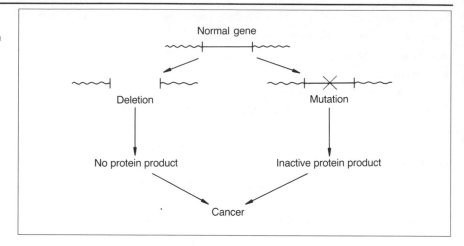

of their mutation or deletion in specific tumor types. Eight of these anti-oncogenes have been cloned and the function of their protein products analyzed (Table 8.3). Most interestingly, the three best-defined anti-oncogene products all encode transcription factors. Two of these, p53 and the Wilms' tumor gene product, appear to act by binding to target sites in the DNA of specific genes and regulating their expression whilst the third, the retinoblastoma gene product (Rb-l), acts predominantly via protein–protein interactions with other regulatory transcription factors. The p53 and Rb-l proteins will therefore be discussed as examples of these two types (for review of the Wilms' tumor gene product, see Hastie, 1993).

Table 8.3 Anti-oncogenes and their functions

Anti-oncogene	Tumors in which gene is mutated	Nature of protein product
APC	Colon carcinoma	Cytoplasmic protein
DCC	Colon carcinoma	Cell adhesion molecule
NF1	Neurofibromatosis	Activator of Ras GTPase activity
NF2	Schwannomas, meningiomas	Cytoplasmic protein
p53	Sarcomas, breast carcinomas, leukemia etc	Transcription factor
RB1	Retinoblastoma, osteosarcoma, small lung cell carcinoma	Transcription factor
VHL	Pheochromocytoma, kidney carcinoma	Membrane protein
WT1	Wilm's tumor	Transcription factor

8.5.2 p53

The p53 protein was originally identified as a 53 kDa protein which bound to the large T oncogene protein of the small DNA virus SV40. Subsequent studies have shown that the gene encoding this protein is mutated in a very wide variety of human tumors, especially carcinomas (for review, see Levine *et al.*, 1991). Most interestingly, inactivation of the p53 gene in mice does not prevent the normal development of the animal. Rather it results in an abnormally high rate of tumor formation which results in the early death of the animal. This has led to the idea that the p53 gene product normally acts to inhibit the development of tumors which occur at high frequency when it is inactivated by mutation (for review, see Berns, 1994).

The detailed characterization of the p53 gene product (for review, see Vogelstein and Kinzler, 1992) has shown that it is a transcription factor capable of binding to a specific DNA sequence and activating the expression of specific genes. The mutations which occur in human tumors result in a loss of the ability to bind to DNA, indicating that the ability of p53 to do this is crucial for its ability to control cellular growth and suppress cancer. The p53 protein thus functions, at least in part, by activating the expression of specific genes whose protein products act to inhibit cellular growth (Figure 8.22i). Its inactivation by gene deletion (Figure 8.22ii) or by mutation (Figure 8.22iii) thus leads to a failure to express these genes, leading to uncontrolled growth.

Interestingly, a failure of p53-mediated gene activation can also occur even in the presence of functional p53 (Figure 8.22iv). Thus many human soft tissue sarcomas contain intact p53 protein but have amplified the cellular *mdm2* oncogene. The MDM2 oncoprotein binds to p53 and masks its activation domain, so preventing it from activating transcription and hence resulting in cancer (Oliner *et al.*, 1993). Such a mechanism parallels the role of the yeast GAL80 protein in masking the GAL4 activation domain which was discussed in Chapter 7 (Section 7.3.4).

Such an interaction of p53 with an oncogenic protein is not unique to the MDM2 oncoprotein. Thus as noted above it was the interaction of p53 with the SV40 large T oncoprotein which led to the original identification of the p53 protein. As in the case of the MDM2 protein, the interaction of p53 with either the large T protein or the transforming proteins of several other DNA viruses prevents p53 from activating its target genes. Such functional inactivation of p53 appears to play a critical role in the ability of these viruses to transform cells to a cancerous phenotype. When taken together with the action of the MDM2 protein this indicates that functional interactions between oncogene and anti-oncogene products are

Figure 8.22 The functional p53 protein acts to stimulate the transcription of genes (GIG) whose protein products inhibit growth (i). This effect can be prevented, however, by the deletion of the p53 gene (ii) or by its inactivation by mutation (iii) as well as by the MDM2 oncoprotein which binds to p53 and masks its activation domain (iv). The activation domain of p53 is indicated by the hatched lines.

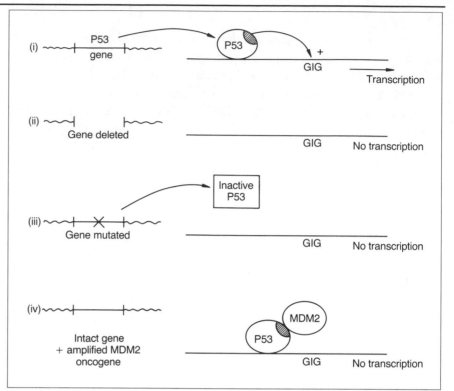

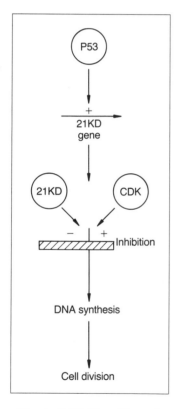

Figure 8.23 The p53 protein stimulates the transcription of the gene encoding the 21 kDa inhibitor of cyclin-dependent kinases (CDK). By inhibiting the activity of the kinases, the 21 kDa protein prevents DNA synthesis and thus cell division.

likely to be critical in the control of cellular growth. Hence changes in this balance due to the overexpression of specific oncogenes or the loss of anti-oncogenes will result in cancer.

The critical role of p53 in regulating the expression of genes encoding growth inhibitory proteins focuses attention on the nature of these growth inhibitory genes. Recently the first such p53 target gene has been identified and shown to encode a 21 kDa protein. When the sequence of the gene encoding this protein was examined it was shown to be identical to a gene cloned by another group which encoded a protein that acted as an inhibitor of cyclin-dependent kinases (for review, see Marx, 1993). Cyclin-dependent kinases are enzymes that stimulate cells to enter cell division. Hence the identification of a p53-regulated gene as an inhibitor of these enzymes immediately suggested that p53 acts by stimulating the expression of this inhibitory factor which in turn inhibits the cyclin-dependent kinases and therefore prevents cells replicating their DNA and undergoing cell division (Figure 8.23).

Hence the p53 gene product plays a key role in regulating cell division, acting by regulating the expression of specific target genes. Its inactivation by mutation or by specific oncogene products is likely to play a critical role in the majority of human cancers.

8.5.3 The retinoblastoma protein

The retinoblastoma gene (Rb-l) was the first anti-oncogene to be identified on the basis that its inactivation results in the formation of eye tumors known as retinoblastomas (for reviews, see Cobrinick *et al.*, 1992; Hamel *et al.*, 1992). As with p53, the Rb-l protein acts as an anti-oncogene by regulating the expression of specific target genes. Unlike p53, however, it appears to act primarily via protein–protein interactions with other transcription factors.

In particular, Rb-l has been shown to interact with the cellular transcription factor E2F. E2F normally stimulates the transcription of several growth-promoting genes such as the cellular oncogenes c-*myc* and c-*myb* (see Section 8.4.3) and the genes encoding DNA polymerase α and thymidine kinase. The interaction of Rb-l with E2F does not affect the ability of E2F to bind to its target sites in the DNA but prevents it stimulating transcription (Figure 8.24). Hence Rb-l acts as an anti-oncogene by preventing the transcription of several growth-promoting genes, including oncogenes, which themselves encode transcription factors.

During the normal cell cycle of dividing cells, the Rb-l protein becomes phosphorylated and this prevents its interacting with E2F, allowing the E2F factor to activate the growth-promoting genes whose protein products are necessary for cell cycle progression (Figure 8.24a). Such an effect can also be achieved by the deletion of the Rb-l gene or its inactivation by mutation which prevents the production of a functional Rb-l protein (Figure 8.24b). Similarly, it is also possible for such an absence of functional Rb-l protein to arise from transcriptional inactivation of the Rb-l gene which is otherwise intact. Thus six out of 77 retinoblastomas were found to have a heavily methylated Rb-l gene which would result in a

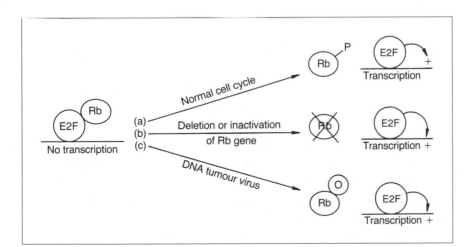

Figure 8.24 The retinoblastoma protein (Rb) binds to the E2F transcription factor and prevents it from activating transcription. This inhibition can be relieved, allowing E2F to activate transcription when (a) the Rb protein is phosphorylated, which occurs during the normal cell cycle and prevents it interacting with E2F, (b) the gene encoding Rb is deleted or inactivated by mutation or (c) the Rb protein binds to the product of a DNA tumor virus oncogene (o) which releases it from E2F.

failure of its transcription (see Section 5.5.1) even though the gene itself was theoretically capable of encoding a functional protein. In addition, as with the p53 protein the Rb-1 protein can be inactivated by protein–protein interaction with the oncogene products of DNA viruses. In this case, however, the association of the viral protein with Rb-1 dissociates the Rb-1–E2F complex, releasing free E2F which can then activate gene expression (Figure 8.24c).

Thus the Rb-1 protein plays a critical role in cellular growth regulation, modulating the expression of specific oncogenes and acting as a target for the transforming oncoproteins of specific viruses. The similarly important role of the p53 protein indicates that, although discovered later than the oncogenes, the anti-oncogenes are likely to play as critical a role in regulating cellular growth in general and the pattern of gene expression, in particular. Hence the precise rate of cellular growth is likely to be controlled by the balance between interacting oncogene and anti-oncogene products, with cancer resulting from a change in this balance by activation or overexpression of oncogenes or inactivation of anti-oncogenes.

8.6 Conclusions

The study of the processes whereby increased expression or mutation of certain cellular genes or the deletion or inactivation of other genes can cause cancer has increased greatly our knowledge of the process of transformation, whereby normal cells become cancerous. Similarly, the recognition that the products of these cellular genes play a critical role in the growth regulation of normal cells has allowed insights obtained from studies of their activities in tumor cells to be applied to the study of normal cellular growth control.

Such a reciprocal exchange of information is well illustrated in the case of the oncogenes and anti-oncogenes that encode cellular transcription factors. Thus, the study of the Rb-1 and p53 proteins has greatly enhanced our knowledge of the processes regulating the transcription of genes whose protein products enhance or inhibit cellular growth. Similarly, the isolation of the *fos* and *jun* genes in tumorigenic retroviruses has aided the study of the effects of growth factors on normal cells, while the recognition that the v-*erbA* gene product is a truncated form of the thyroid hormone receptor has allowed the elucidation of its role in transformation via the inhibition of erythroid differentiation.

These two cases also illustrate the two mechanisms by which cellular genes can become oncogenic. Thus in the case of the v-*erbA* gene, mutations have rendered the protein different from the corresponding c-*erbA* gene from which it was derived. These alter the properties of the protein so it cannot bind thyroid hormone and it behaves as a dominant repressor of transcription. Similar activation by mutation is also seen in oncogenes, where products function at other stages of the growth-signaling process. Notable examples include the truncation of the epidermal growth factor receptor in the v-*erbB* oncogene and the loss of GTP-hydrolyzing ability in the *ras* oncogene, both of which result in proteins 'frozen' into a positive growth-signaling form. Interestingly, the interaction between oncogene and anti-oncogene products encoding transcription factors discussed in Section 8.5 is also paralleled at other stages. Thus the NFl anti-oncogene encodes a protein which activates the GTP-hydrolyzing ability of the *ras* oncogene product, so switching it to the non-growth-promoting form.

In contrast to the situation with oncogenes encoding proteins which pre-exist in an inactive form, oncogenes whose products are made only in response to a particular growth signal and whose activity then mediates cellular growth can cause cancer simply by the normal product being made at an inappropriate time. Thus, in the case of Fos or Jun, which are synthesized in normal cells in response to treatment with growth-promoting phorbol esters or growth factors, their continuous synthesis is sufficient to transform the cell. A similar case is that of the *sis* oncogene, which encodes the cellular gene for platelet-derived growth factor and also causes cancer when overexpressed by a retrovirus.

Such cases of high-level expression of a normal oncogene product causing cancer also occur in a number of cases without any evidence of retroviral involvement. These examples of alterations in cellular regulatory processes producing increased expression of particular genes obviously provide another aspect to the connection between cancer and gene regulation. Thus, information on the origin of the translocations of the c-*myc* gene in Burkitt's lymphoma is obviously important for the study of cancer etiology. Similarly, the fact that such translocations increase expression in some cases by removing elements that normally inhibit c-*myc* expression will allow the characterization of such negative elements and their role in regulating c-*myc* expression in normal cells.

Hence the study of cellular oncogenes has contributed greatly to our knowledge both of cancer and of cellular growth regulatory processes, and is likely to continue to do so in the future.

References

Angel, I. P., M. Imagawa, R. Chiu, B. Stein, R. J. Imbra, J. J. Rahmsdorf, C. Jonat, P. Herlich and M. Karin 1987. Phorbol ester-inducible genes contain common *cis* element recognized by a TPA-modulated *trans*-acting factor. *Cell* **49**, 729–39.

Angel, P., E. A. Allegretto, S. T. Okino, K. Hattori, W. J. Boyle, T. Hunter and M. Karn 1988. Oncogene jun encodes a sequence-specific trans-activator similar to AP-l. *Nature* 332, 166–70.

Berns, A. 1994. Is p53 the only real tumour suppressor gene? *Current Biology* 4, 137–9.

Beug, H., P. Kahn, B. Vennstrom, M. J. Hayman and T. Graf 1985. How do retroviral oncogenes induce transformation in mammalian cells? *Proceedings of the Royal Society of London, Series B* **226**, 121–6.

Bishop, J. M. 1987. The molecular genetics of cancer. *Science* **235**, 305–11.

Blackwood, F. M., L. Kretzner and R. N. Eisenmann 1992 Myc and Max function as a nucleoprotein complex. *Current Opinion in Genetics and Development* 2, 227–35.

Bos, T. J., D. Bohmann, H. Tsuchie, R. Tjian and P. K. Vogt 1988. V-*jun* encodes a nuclear protein with enhancer binding properties of AP-l. *Cell* **52**, 705–12.

Bourne, H. R. and H. E. Varmus (eds) 1992. Oncogenes and cell proliferation. *Current Opinion in Genetics and Development* **2**, 1–57.

Cobrinik, D., S. F. Dowdy, P. W. Hines, S. Mittnacht and R. A. Weinberg 1992. The retinoblastoma protein and the regulation of cell cycling. *Trends in Biochemical Sciences* 17, 312–15.

Collins, S. and M. Groudine 1982. Amplification of endogenous *myc*-related DNA sequences in a human myeloid leukemia cell line. *Nature* **298**, 679–81.

Curran, T. and B. R. Franza 1988. Fos and Jun: the AP-l connection. *Cell* **55**, 395–7.

De Feo-Jones, D., E. M. Scolnick, R. Koller and R. Dhar 1983. *Ras*-related sequences identified and isolated from the yeast *Saccharomyces cerevisciae*. *Nature* **306**, 707–9.

Forrest, D. and T. Curran, 1992. Crossed signals: oncogenic transcription factors. *Current Opinion in Genetics and Development* **2**, 19–27.

Grandchamp, B., C. Picat, V. Mignotte, J. H. P. Wilson, K. TeVelde, L. Sandkuyl, P. H. Romeo, M. Goossens and Y. Nordmann 1989. Tissue-specific splicing mutation in acute intermittent porphyria. *Proceedings of the National Academy of Sciences of the USA* **86**, 661–4.

Graupner, G., K. N. Wills, M. Tzukerman, X.-K. Zhang and M. Pfuli 1989. Dual regulatory role for thyroid-hormone receptors allows control of retinoic acid receptor activity. *Nature* **340**, 653–6.

Halazonetis, T. D., K. Georgopoulos, M. E. Greenberg and P. Leder 1988. C-*jun* dimerizes with itself and with c-fos forming complexes of different DNA binding abilities. *Cell* **55**, 917–24.

Hamel, P.A., B. L. Gallie and R.A. Phillips 1992. The retinoblastoma protein and cell cycle regulation. *Trends in Genetics* **8**, 180–5.

Hastie, N. D. 1993. Wilms' tumour gene and function. *Current Opinion in Genetics and Development* **3**, 408–13.

Hayward, W. S., B. G. Neel and S. M. Astin 1981. Activation of a cellular onc gene by promoter insertion in ALV-induced Lymphoid leukosis. *Nature* **290**, 475–80.

Knudson, A. G. 1993. Antioncogenes and human cancer. *Proceedings of the National Academy of Sciences of the USA* **90**, 10914–21.

Lamph, W. W., P. Wamsley, P. Sassono-Corsi and I. M. C. Verma 1988. Induction of proto-oncogene Jun/Ap-l by serum and TPA. *Nature* **334**, 629–31.

Lee, W., P. Mitchell and R. Tjian 1987. Purified transcription factor AP-l interacts with TPA-inducible enhancer elements. *Cell* **49**, 741–52.

Levine, A.J., J. Momaid and C. A. Finlay 1991. The p53 tumor suppressor gene. *Nature* **351**, 453–6.

Lewin, B. 1991. Oncogenic conversion by regulatory changes in transcription factors. *Cell* **64**, 303–12.

Marth, J. D., R. W. Overell, K. E. Meier, E. G. Krebs and R. M. Perlmutter 1988. Translational activation of the Lck proto-oncogene. *Nature* **332**, 171–3.

Marx, J. 1993. How p53 suppresses cell growth. *Science* **262**, 1644–5.

Moran, E. 1993. DNA tumour virus transforming proteins and the cell cycle. *Current Opinion in Genetics and Development* **3**, 63–70.

Oliner, J. B., J. A. Pietenpol, S. Thiagalingam, J. Gyuris, K. W. Kinzler and B. Vogelstein, 1993. Oncoprotein MDM2 conceals the activation domain of tumour suppressor p53. *Nature* **362**, 857–60.

Radovick, S., M. Nations, Y. Du, L. A. Berg, R. D. Weintraub and F. E. Wondisford, 1992. A mutation in the POU homeodomain of Pit-l responsible for combined pituitary hormone deficiency. *Science* **257**, 1115–18.

Ransone, L. J. and I. M. Verma, 1990. Nuclear proto-oncogenes: Fos and Jun. *Annual Review of Cell Biology* **6**, 531–57.

Rauscher, F. J., D. R. Cohen, T. Curran, T. J. Bos, P. K. Vogt, D. Bohmann, R. Tjian and B. R. Franza 1988. Fos-associated protein P39 is the product of the c-*jun* oncogene. *Science* **240**, 1010–16.

Reith, W., S. Satola, C. H. Sanchey, I. Amaldi, B. Lisowska-Grospiere, C. Griscelli, M. R. Hadam and B. Much 1988. Congenital immunodeficiency with a regulatory defect in MHC Class II gene expression lacks a specific HLA-DR promoter binding protein RF-X. *Cell* **53**, 897–906.

Sap, J., A. Munoz, K. Damm, Y. Goldberg, J. Ghysdael, A. Leutz, H. Beug and B. Vennstrom 1986. The c-*erb*-A protein is a high-affinity receptor for thyroid hormone. *Nature* **324**, 635–40.

Sap, J., A. Munoz, A. Schmitt, H. Stunnenberg and B. Vennstrom 1989.

Repression of transcription at a thyroid hormone response element by the v-*erbA* oncogene product. *Nature* **340**, 242–4.

Schuermann, M., M. Neuberg, J. B. Hunter, T. Jenuwein, R.-P. Ryseck, R. Bravo and R. Muller 1989. The leucine repeat motif in *fos* protein mediates complex formation with Jun/Apl and is required for transformation. *Cell* **56**, 507–16.

Schule, R. and R. M. Evans 1991. Cross coupling of signal transduction pathways: zinc finger meets leucine zipper. *Trends in Genetics* **7**, 377–81.

Spencer, C. A. and M. Groudine 1991. Control of c-*myc* regulation in normal and neoplastic cells. *Advances in Cancer Research* **56**, 1–48.

Takeya, T. and H. Hanafusa 1985. Structure and sequence of the cellular gene homologous to the RSV-*src* gene and the mechanism for generating transforming virus. *Cell* **32**, 881–90.

Trottier, Y., D. Dewys and J. L. Mandel 1993. Fragile X syndrome: an expanding story. *Current Biology* **3**, 783–6.

Vogelstein, B. and K. Kinzler 1992. p53 function and dysfunction. *Cell* **70**, 523–6.

Vogt, P. K., T. J. Bos and R. F. Doolittle 1987. Homology between the DNA binding domain of the GCN4 regulator protein of yeast and the carboxylterminal region of a protein coded for by the oncogene *jun*. *Proceedings of the National Academy of Sciences of the USA* **84**, 3316–19.

Weinberg, R. A. 1985. The action of oncogenes in the cytoplasm and nucleus. *Science* **230**, 770–6.

Weinberg, R. A. 1993. Tumour suppressor genes. *Neuron* **11**, 191–6.

Weinberger, C., C. C. Thompson, E. S. Ong, R. Lebo, D. J. Gruol and R. M. Evans 1986. The c-*erb-A* gene encodes a thyroid hormone receptor. *Nature* **324**, 641–6.

Zenke, M., D. Kahn, C. Disela, B. Bennstrom, A. Leutz, K. Keegan, M. J. Hayman, H. R. Chui, N. Yew, J. D. Engel and H. Berg 1988. V-*erb-A* specifically suppresses transcription of the avian erythrocyte anion transporter (Band 3) gene. *Cell* **52**, 107–19.

Conclusions and future prospects

9

The extraordinary rate of progress in the study of gene regulation can be gauged from the fact that by 1980, no transcriptional regulatory protein or its DNA-binding site in a regulated promoter had been defined. The tremendous advances since that time were evident by the time the first edition of this book was published. Thus at that time (1990) a relatively clear picture of the action of individual factors which regulate gene expression was available. For example, in the case of gene induction by glucocorticoid hormone it was known that these agents act by activating specific receptor proteins (Section 7.4.3), that such activated receptors bind to specific DNA sequences upstream of the target gene (Section 6.2.3), displacing a nucleosome (Section 5.6.2) and that a region in the receptor protein then interacts with a factor bound to the TATA box to activate transcription (Section 7.3.3).

Clearly, therefore, the process by which a single agent can activate a specific transcription factor and thereby modulate gene expression was reasonably well understood in outline. However, in the period from 1990 to the present day, it has become increasingly clear that the activation of a single transcription factor by a single agent cannot be assessed in isolation. Rather the activity of a specific factor will depend upon its interaction with other transcription factors as well as with other proteins. Thus, for example, glucocorticoid hormone activates transcription by promoting the dissociation of the glucocorticoid receptor from the inhibitory protein hsp90 which otherwise anchors it in the cytoplasm and prevents it from moving to the nucleus and activating transcription (Section 7.4.3).

A similar example of protein–protein interactions regulating transcription factor activity but in the opposite direction is observed in the Fos and Jun proteins which regulate gene expression in response to phorbol ester treatment (Section 8.4.1). Thus, as noted in Chapter 7 (Section 7.2.4), the Fos protein cannot bind to DNA on its own but can do so only after forming a heterodimer with the Jun protein. This heterodimer then binds to DNA and activates transcription. Hence in this case, the interaction with another protein allows Fos to activate transcription whereas in the case of the glucocorticoid receptor the interaction with hsp90 has the opposite effect.

Such types of protein–protein interactions regulating transcription factor activation are not confined to the regulation of gene expression by short-term inducers but are also involved in the regulation of tissue-specific gene expression. Thus as discussed in Chapter 7 (Section 7.2.4) the MyoD transcription factor can bind to the promoters of muscle-specific genes and activate their expression, thereby causing the differentiation of a fibroblast cell line into muscle cells. The production of differentiated skeletal muscle cells in this situation requires, however, that the cells are cultured in the absence of growth factors. In the presence of growth factor, the cells contain high levels of the inhibitory factor Id which heterodimerizes with MyoD and prevents it binding to DNA and activating transcription. The production of the skeletal muscle phenotype therefore requires not only high levels of the activating MyoD protein but also a fall in the levels of the inhibitory Id protein which is induced by growth factor removal.

It is clear therefore that in families of transcription factors which are capable of dimerization by means of specific motifs such as the leucine zipper or helix-loop-helix (see Section 7.2.4), such dimerization with another member of the family can result in activation (Fos–Jun) or repression (MyoD–Id). The situation is further complicated, however, by the fact that interaction can also occur between members of different families which are mediating the response to different cellular signals. Thus as discussed in Chapter 8 (Section 8.4.1) the Fos–Jun dimer can interact with the activated glucocorticoid receptor to form a complex that cannot bind either to the AP1 DNA binding site for Fos–Jun or to the glucocorticoid response element. Hence such an interaction results in a mutual inhibition of gene activation by each of these factors.

This example illustrates therefore how the effect of one signaling pathway can be affected by the activation of another signaling pathway. Hence in the absence of glucocorticoid hormone, the glucocorticoid receptor will be anchored in the cytoplasm by hsp90 and the Fos–Jun complex will be able to activate gene expression in response to growth factors or phorbol esters. However, when the glucocorticoid receptor is freed from hsp90 by the presence of glucocorticoid it can interact with Fos–Jun and prevent them activating gene expression. This is seen in the case of the gene encoding the collagenase enzyme which is activated by the Fos–Jun complex in response to phorbol esters and can produce severe tissue destruction in an inflamed area. As expected on the basis of the above model, the activation of this gene by phorbol esters is inhibited by treatment with glucocorticoid hormone, accounting for the anti-inflammatory effect of this steroid hormone. Conversely, the activation of glucocorticoid-responsive genes by hormone treatment will be inhibited by the presence of high levels of Fos and Jun induced by growth factor or phorbol ester treatment.

These examples also illustrate another theme which has become of increasing importance in gene regulation, i.e. the importance of specific inhibition as opposed to activation of gene expression (see Section 7.3.4). This can be achieved both by transcription factors such as the Id factor which function only as inhibitors or by factors such as Fos–Jun or the glucocorticoid receptor which can function both as activators or as repressors depending on the other factors which are present. Indeed in some cases the same factor can also function as a direct activator or repressor in the absence of any other transcription factor. This is seen in the case of the thyroid hormone receptor which unlike the glucocorticoid receptor binds to its DNA target site even in the absence of hormone. As discussed in Chapter 7 (Section 7.3.4) this factor can directly inhibit promoter activity in the absence of thyroid hormone. In the presence of thyroid hormone, however, it undergoes a conformational change which exposes its activation domain and allows it to activate gene expression.

Interestingly, the same ability to act as an activator or repressor following DNA binding is seen in the case of the glucocorticoid receptor. In this case, however, this transition is not controlled by the presence or absence of the hormone but by the nature of the target site to which the receptor binds following hormone treatment and dissociation from hsp90. In most cases, binding of the receptor to a glucocorticoid response element (GRE; Section 6.2.3) allows it to activate transcription and hence results in the stimulation of gene expression in response to hormone treatment. In contrast, however, some genes such as that encoding prolactin contain a target site distinct from the GRE which is still capable of binding receptor. The binding of the activated receptor to this so called nGRE results in inhibition of gene expression in response to glucocorticoid hormone, presumably because the receptor binds to the nGRE in a distinct configuration in which its activation domain is not exposed. Hence the glucocorticoid receptor can repress gene expression by two distinct mechanisms involving either binding to an nGRE or formation of a complex with Fos–Jun which prevents it binding to DNA. Both these forms of repression are dependent on treatment with the hormone since the glucocorticoid receptor is bound to hsp90 in the absence of the appropriate hormone and hence cannot bind to DNA or to Fos–Jun until released from hsp90 by hormone treatment.

These examples indicate therefore that the effect of a particular factor on gene expression will depend on the nature of the other factors present in the cell, the state of activation of specific signaling pathways and the DNA binding site to which it binds. Moreover, the effect of a specific binding site and the factor(s) which bind to it will clearly depend on its relationship to the other DNA binding sites in the promoter of a specific gene as well as the nature of those sites and the proteins which bind to them.

Over the last few years it has therefore become clear that the regulation of a single gene in an intact cell is vastly more complex than could be predicted from the study of an isolated transcription factor interacting with its binding site. At first sight therefore it may appear that recent progress in this area has merely complicated the issue. However, the process of gene regulation must not only produce all the different types of cells in the body but must ensure that each cell type is produced in the right place and at the right time during development. It is inevitable therefore that gene regulation must be a highly complex process with numerous factors interacting with each other and their respective DNA binding sites to activate or inhibit the expression of specific genes in specific cell types. The progress in understanding these effects in the last few years offers hope therefore that an understanding of mammalian development in terms of differential gene expression can ultimately be achieved.

Index